Lecture Notes in Mathematics

A collection of informal reports and seminars
Edited by A. Dold, Heidelberg and B. Eckmann, Zürich

165

Joel M. Cohen

University of Pennsylvania, Philadelphia, PA/USA

Stable Homotopy

Springer-Verlag
Berlin · Heidelberg · New York 1970

Offsetdruck: Julius Beltz, Weinheim/Bergstr.

Introduction

These notes are essentially the lecture notes of a course I gave at the University of Chicago in the summer of 1968.* Most aspects of stable homotopy are touched on and some are studied in very great detail. It should, however, be emphasized that we are only concerned with finite CW complexes. Thus one never has to worry about the problems which may arise for infinite CW complexes; i.e. certain long exact sequences which are easy to get for finite dimensional CW complexes become very difficult in general unless one takes great care in defining the morphisms (as J. M. Boardman has done in his Warwick lecture notes; or see Tierney).

It is assumed that the reader has had a year of algebraic topology (a course which covers the equivalent of most of Spanier, say). I quote without proof some theorems from first year topology (e.g. the Hurewicz theorem) and prove others. In addition I assume the reader has some understanding of spectral sequences and what they can do. Specifically, I assume existence of the Serre Spectral Sequence in homology. Spanier covers quite adequately the necessary material.

For the computations of the stable homotopy groups of spheres in Chapter V, I quote a lot of results on the Steenrod Algebra-- all of which can be found in Steenrod-Epstein or Mosher-Tangora. Lack of prior knowledge of cohomology operations will not interfere with the understanding of this section, although the reader may have to accept some results on faith (or study the above-mentioned books).

This set of notes has a quite different point of view on the whole from Frank Adams' lecture notes on stable homology. I feel

*The author was partially supported by the National Science Foundation during the preparation of these notes.

that to some degree, these complement the other. Although I do construct the Adams spectral sequence for completeness, not very much is said about it here and the reader is encouraged to pursue the subject either in Adams' notes or in Mosher-Tangora. The present method of computing the stable homotopy groups of spheres is somewhat simpler than the Adams spectral sequence in the dimensions where it is done. (Higher up this method seems to break down and the Adams method is much neater.)

Chapter IV, on stable homotopy and category theory is entirely the work of Peter Freyd. The proofs are to some extent my own--I tried to make them more topological than category theoretical where possible; but the fact remains that the main results, which are purely topological statements, cannot be proved without using (or directly mimicking) Freyd's embedding of the stable homotopy category into an abelian category.

Thanks are due many people for the ideas incorporated in these notes. My interest in the subject was aroused by George Whitehead; much of my thinking was influenced by him and several proofs are lifted directly from him. Chapter V is an abridged version of my thesis written under Donald Anderson. I express my deep gratitude to him for many helpful suggestions during the original writing and since. In addition many parts of these notes grew out of very useful discussions with Frank Peterson, David Kraines, Gerald Porter, Peter May, Peter Freyd and Brayton Gray. I wish to thank Susan McMahon, Mary Vallery and Cecelia Ricciotti for putting up with my handwriting and typing this manuscript.

TABLE OF CONTENTS

CHAPTER 0. PRELIMINARIES

To start with, we shall consider based, simply-connected spaces; i.e., every space X comes equipped with a basepoint $* \in X$, X is connected and $\pi_1(X,*) = 0$. $f:X \to Y$ will always be a continuous map with $f(*) = *$. $F(X,Y)$ is the set of all such maps. We give $F(X,Y)$ the compact-open topology. Let $\mathcal{S}$ be this category.

A homotopy from X to Y is a continuous path H_t in $F(X,Y)$, $0 \le t \le 1$. Given such a path we say $f \sim g$, f is homotopic to g for $f = H_o$, $g = H_1$. (Observe that $H_t(*) = *$ for all t.) The set of homotopy classes $[f]$ of maps $f:X \to Y$ is $\pi_o(F(X,Y)) = [X,Y]$. $F(X,Y)$ has a basepoint $*$ where $*(x) = *$ for all $x \in X$.

<u>Notation</u>: Two spaces X and Y are homotopy equivalent, $X \simeq Y$, if and only if there are maps $f:X \to Y$, $g:Y \to X$ with $fg \sim 1_Y$, $gf \sim 1_X$. We shall write $X \cong Y$ if X and Y are homeomorphic. (I.e., $fg = 1_Y$, $gf = 1_X$ for some f and g.)

If A is a subspace of X containing the basepoint of X as its own (this is necessary, of course, in order to have the inclusion map basepoint preserving) then X/A is X with A identified to the basepoint. If it should happen that X has no basepoint and $A \subset X$, then X/A still makes sense and now has a basepoint.

Given spaces X and Y we form the <u>wedge</u> (essentially the one point union) $X \vee Y = X \times * \cup * \times Y \subset X \times Y$, with $* \times *$ as basepoint. Then we define the "<u>smash product</u>" or <u>reduced join</u> $X \wedge Y = \dfrac{X \times Y}{X \vee Y}$. $\vee$ and $\wedge$ are commutative bifunctors $\mathcal{S} \times \mathcal{S} \to \mathcal{S}$ and $\wedge$ distributes over $\vee$.

Observe that $\times$ is the product and $\vee$ the coproduct for all maps and also for homotopy classes; i.e.,

$F(X, Y \times Z) \cong F(X,Y) \times F(X,Z)$, $F(X \vee Y, Z) \cong F(X,Z) \times F(Y,Z)$,

$[X, Y \times Z] = [X,Y] \times [X,Z]$, and $[X \vee Y, Z] = [X,Z] \times [Y,Z]$.

Let $\rho : F(X \wedge Y, Z) \to F(X, F(Y,Z))$ be given by $[\rho(f)(x)](y) = f(x \wedge y)$. ρ is continuous. If Y is locally compact, then ρ is, in fact, a homeomorphism. Thus, for Y locally compact, the functors $- \wedge Y$ and $F(Y,-)$ are adjoint (or by commutativity, $Y \wedge -$ and $F(Y,-)$) .

We define the 1-sphere $S^1 = \frac{I}{\{0,1\}}$ where I = the unit interval $[0,1]$ with basepoint 0 . Let $S = S^1 \wedge -$ a functor $\mathcal{S} \to \mathcal{S}$. By the above, it has a right adjoint $\Omega = F(S^1,-)$. $F(SX,Z) = F(X,\Omega Z)$ so $[SX,Z] = [X,\Omega Z]$. We recall that $[SX,Z]$ has a group structure arising from the "pinch" map $S^1 \to S^1 \vee S^1$.

If n is an integer define a function $n: I \to I$ by $n(t) = n \cdot t - [nt]$ ($[x]$ = greatest integer $\leq x$). n is not continuous but its composition with the projection $I \to S^1$ is. Since $n(\{0,1\}) = * \in S^1$ we in fact have $n: S^1 \to S^1$ defined. (The reader will have to excuse the fact that n represents three different things.) Considering S^1 as $\{\text{complex } z \mid |z| = 1\}$, $n(z) = z^n$.

For each integer $r > 1$ we can define $S^{r+1} = S(S^r)$. In fact since $S^1 = S(\{0,1\})$ we see that $\{0,1\}$ is a good choice for S^0 - the zero sphere. S^r will also represent the functor $S^r \wedge -$. If n is an integer $S^{r-1}(n): S^r \to S^r$ is defined and for confusion will be written as $n: S^r \to S^r$.

There is one more functor that we wish to consider: Let $\tilde{I}$ be I with the basepoint 1 . Then the cone functor T is $\tilde{I} \wedge -$. We shall embed $X \subset TX$ by $x \mapsto (0,x)$. Then we see that a space X may be contracted to one point if and only if X is a retract of TX ; and a map $f:X \to Y$ is null homotopic if and only if it can be extended to a map $TX \to Y$.

We recall the following basic result (cf. Spanier for example):

Lemma 0.1: If $f:S^r \to S^r$ then for some integer n , $f \sim n$ and if $m \neq n$, $m \not\sim n$. In other words, $[S^r,S^r] \cong Z$, the integers, with the identity corresponding to 1 . (Here of course $r \geq 1$.)

■*

Recall the homotopy group functor, $\pi_n = [S^n, \ ]$. We recall that a non-representable version can be defined on pairs. Now $H_n(S^n) \cong Z$. For each n , choose $\iota_n \in H_n(S^n)$ a generator, so that ι_{n+1} corresponds to ι_n under the natural isomorphism $H_n(S^n) \cong H_{n+1}(S^{n+1})$. This defines the Hurewicz map, a natural transformation, $h_n:\pi_n \to H_n$: if $[f] = \alpha \in \pi_n(X)$ then $h_n(\alpha) = f_*(\iota_n) \in H_n(X)$. For pairs of spaces satisfying a certain property, and $n > 0$, $H_n(X,A) \cong H_n(X/A,*)$. Then we have defined $h_n(X,A):\pi_n(X,A) \to H_n(X,A)$ as the composite, for $n > 0$,

$$\pi_n(X,A) \to \pi_n(X/A,*) = \pi_n(X/A) \xrightarrow{h_n(X/A)} H_n(X/A) \cong H_n(X,A)$$

The main theorem involving $h_n(X,A)$ is

Theorem 0.2 (Hurewicz): If $\pi_i(X,A) = 0$, $0 \leq i < n$ then $h_n(X,A)$ is an isomorphism.

---

* This symbol will be used to indicate that no further proof of the theorem will be given.

It is important to note that we must assume X and A simply connected.

The "condition" referred to above is this:

<u>Definition</u>: (X,A) has the homotopy extension property (HEP) if $A \subset X$, and $I \times A \cup 0 \times X$ is a retract of $I \times X$.

Observe that if (X,A) has the HEP then letting $r: I \times X \to I \times A \cup 0 \times X$ be the retraction we have $r_1 = r(1,-): X \to I \times A \cup 0 \times X$ with $r_1(A) = 1 \times A$. Observe that for $X \cup TA$ (this always means that $a = (0,a)$) we have $X \cup TA = \dfrac{I \times A \cup 0 \times X}{1 \times A}$. Since $r_1(A) = 1 \times A$, r_1 induces a map $f: X/A \to X \cup TA$. Defining $g: X \cup TA \to X/A$ by $g(x) = [x]$, $g(t,a) = *$, we observe that g is the homotopy inverse to f.

Thus

<u>Theorem 0.3</u>: If (X,A) satisfies HEP, then $f: X/A \simeq X \cup TA$. Furthermore $SA = \dfrac{X \cup TA}{X}$, composing f with the projection yields $p: X/A \to SA$ called the canonical map. p is unique up to homotopy and is natural once $r: I \times X \to I \times A \cup 0 \times X$ is given.

<u>Remark</u>: 1. The HEP for a pair (X,A) is satisfied if and only if the following is true: given $f: X \to Y$ and a homotopy $H: A \times I \to Y$ beginning at $f|A$, then there exists a homotopy $G: X \times I \to Y$ beginning at f and with $G|A = H$.

2. The HEP always holds for a pair $(A \cup e^n, A)$ where $\text{int}\, e^n \cap A = \emptyset$ and $\text{int}\, e^n \cong \text{int}\, I^n$; i.e. e^n is an attached n-cell. See <u>Hu</u> [2].

CHAPTER 1. HOMOTOPY AND HOMOLOGY NOT-SO-LONG EXACT SEQUENCES

1.1 Basic Properties of Mapping Cones

If $f:X \to Y$ we define the _mapping cylinder_ $Z_f = Y \cup I \times X/\sim$ where $(0,x) \sim f(x) \in Y$ and $(t,*) \sim *$. $Y \subset Z_f$ is a strong deformation retract, hence a homotopy equivalence. We include X in Z_f, $i:X \hookrightarrow Z_f$ by $i(x) = (x,1)$. Then the following diagram homotopy commutes:

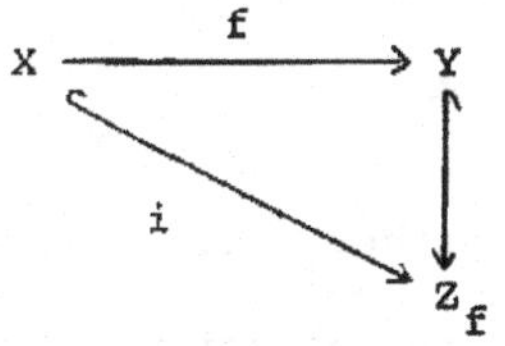

Thus in homotopy theory we may take any map to be an inclusion by replacing the codomain by a homotopy equivalent space and the map by a homotopic one.

We now define the _mapping cone_ or _cofibre_ of f, $C_f = Z_f/X = Y \cup TX/\sim$ with $(0,x) \sim f(x)$. Then we have canonically the inclusion $i_f:Y \to C_f$ and the projection $\sigma_f:C_f \to C_f/Y = SX$.

We now prove several basic properties.

1) If $f \simeq g:X \to Y$ then $C_f \simeq C_g$: let $H:X \times I \to Y$ be such that $H(x,0) = f(x)$, $H(x,1) = g(x)$. Then define $\varphi:C_f \to C_g$ by $\varphi(y) = y$ and

$$\varphi(t,x) = \begin{cases} H(x,2t) & 0 \leq t \leq \frac{1}{2} \\ (2t-1,x) & \frac{1}{2} \leq t \leq 1 \end{cases}$$

. This makes sense since $\varphi(0,x) = H(x,0) = f(x)$ and $(0,x) = \varphi(\frac{1}{2},x) = H(x,1) = g(x)$.

Similarly we define $\psi:C_g \to C_f$ and it is easy to show that

ψ is the homotopy inverse of φ.

2) Let $a:X \to X'$ be a homotopy equivalence with $\hat{a}$ its homotopy inverse. Let $f:X' \to Y$. Then there is a map $\varphi:C_{fa} \to C_f$ by $\varphi(y) = y$, $\varphi(t,x) = (t,a(x))$. Similarly there is a map $C_{fa\hat{a}} \to C_{fa}$. But by 1), $C_{fa\hat{a}} \simeq C_f$ since $a\hat{a} \sim 1_{X'}$. Thus there are maps $C_f \leftrightarrows C_{fa}$ and it is not difficult to show that they are homotopy inverses. Similarly, if $g:Z \to X$, then $C_{ag} \simeq C_g$.

3) By Theorem 0.3 we have: if $f:Y \to X$ is an inclusion and (X,Y) has the HEP, then $C_f \simeq X/Y$.

4) Putting this all together, this says that, up to homotopy type, we may replace a map by an inclusion and the cone by the quotient space in order to study the mapping cone sequence. For example

5) If $f:X \to Y$ and $g:Y \to Z$ are maps, then $\varphi:C_f \to C_{gf}$ is defined by $\varphi(y) = g(h)$, $\varphi(t,x) = (t,x)$. Then $C_\varphi \simeq C_g$: assume f and g are inclusions having the HEP. Then φ is also and $C_g \simeq Z/Y = (Z/X)/(Y/X) \simeq C_{gf}/C_f \simeq C_\varphi$.

6) If $f:X \to Y$ and Z is any space, then $[C_f,Z] \xrightarrow{i_f^*} [Y,Z] \xrightarrow{f^*} [X,Z]$ is exact: if $H:X \times I \to C_f$ by $H(t,x) = (t,x)$ then $H(x,0) = (i_f \circ f)(x)$ and $H(x,1) = *$. Thus $i_f \circ f \sim *$. So $f^* \circ i_f^* = 0$. Conversely, if $g:Y \to Z$ and $g \circ f \sim *$, let $G:X \times I \to Z$ be such that $G(x,1) = *$ and $G(x,0) = g \circ f$. Then define $\tilde{g}:C_f \to Z$ by $\tilde{g}(y) = g(y)$, $\tilde{g}(t,x) = G(x,t)$. This is well-defined and $\tilde{g} \circ i_f = g$.

7) (C_f,Y) has the HEP so that $\theta:C_{i_f} \simeq SX$. The following then are

homotopy commutative diagrams:

$$\begin{array}{ccc} C_f & \hookrightarrow & C_{i_f} \\ & {\scriptstyle \sigma_f}\searrow & \downarrow{\scriptstyle \theta} \\ & & SX \end{array} \qquad \begin{array}{ccc} C_{i_f} & \xrightarrow{\sigma_{i_f}} & SY \\ {\scriptstyle \theta}\downarrow & \nearrow{\scriptstyle \pm Sf} & \\ SX & & \end{array}$$

(the sign depends on the actual choice of θ , but usually will come out $-$).

8) If $f:X \to Y$ and Z is some space, then form $f \wedge 1_Z : X \wedge Z \to Y \wedge Z$. Then there is a natural map $C_{f\wedge 1_Z} \to C_f \wedge Z$ which is a bijection: take f to be an inclusion; then we get the map $Y\wedge Z/X\wedge Z \to (Y/X) \wedge Z$. For Y and X compact, it is a homeomorphism. In general, it induces an isomorphism of homotopy groups.

By a cofibration or a mapping cone sequence, we mean a sequence $X \xrightarrow{f} Y \xrightarrow{g} Z$, such that there is a homotopy equivalence $a:Z \to C_f$ and $a\circ g \simeq i_f : Y \to C_f$. Thus if (X,A) has the HEP, $A \to X \to X/A$ is a mapping cone sequence.

We get the Barratt-Puppe sequence from the above constructions:

$$(\alpha) \qquad X \xrightarrow{f} Y \xrightarrow{i_f} C_f \xrightarrow{\sigma_f} SX \xrightarrow{Sf} SY \xrightarrow{Si_f} SC_f \to \cdots$$

which has the property that every sequence of two maps (and three spaces) is a mapping cone sequence. Also observe that if W is any space, then $[(\alpha),W]$ is a long exact sequence.

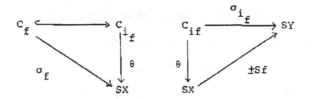

1.2 Basic Properties of Fibres

There is an adjoint construction to that of mapping cone. Set $PY = \{\omega: I \to Y | \omega(0) = *\}$. If $f: X \to Y$ let $E_f = \{(x,\omega) \in X \times PY | \omega(1) = f(x)\}$. E_f is called the <u>fibre</u> of f. We define $j_f: E_f \to X$ by $j_f(x,\omega) = x$ and $\varphi_f: \Omega Y \to E_f$ by $\varphi_f(\omega) = (*,\omega)$.

We have the dual properties:

1') If $f \sim g$ then $E_f \simeq E_g$

2') If $a: X' \to X$ and $b: Y \to Y'$ are homotopy equivalences and $f: X \to Y$, then $E_{b \circ f \circ a} \simeq E_f$.

5') Given $X \xrightarrow{f} Y \xrightarrow{g} Z$, there is a map $\varphi: E_{gf} \to E_g$ given by $\varphi(x,\omega) = (f(x),\omega)$ with fibre $E_\varphi \simeq E_f$.

6') If $f: X \to Y$ and Z is any space, then $[Z,E_f] \xrightarrow{j_{f*}} [Z,X] \xrightarrow{f_*} [Z,Y]$ is exact.

7') There is a homotopy equivalence $\varphi: E_{j_f} \simeq \Omega Y$ with the following homotopy commutative:

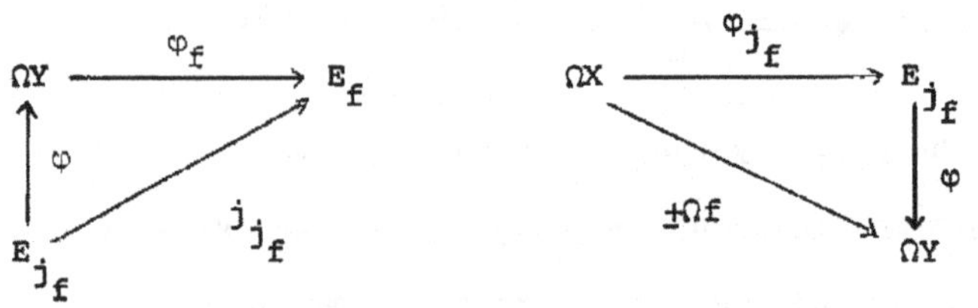

8') If $f: X \to Y$ and Z is some space, then form $f \times 1_Z: X \times Z \to Y \times Z$. Then since $P(Y \times Z) = PY \times PZ$,

$E_{f\times 1_Z} = \{((x,z),(\omega_1,\omega_2)) \in (X \times Z) \times (PY \times PZ) \mid \omega_1(1) = f(x), \omega_2(1) = Z\}$

$= \{(x,\omega) \in X \times PY \mid \omega(1) = f(x)\} \times \{(z,\omega) \in Z \times PZ \mid \omega(1) = z\} \times Z$

$= E_f \times PZ \times Z$. Since PZ is contractible, $E_{f\times 1_Z} \simeq E_f \times Z$.

Finally, there is a Barratt-Puppe sequence

$$(\beta) \qquad \cdots \to \Omega E_f \xrightarrow{\Omega j_f} \Omega X \xrightarrow{\Omega f} \Omega Y \xrightarrow{\sigma_f} E_f \xrightarrow{j_f} X \xrightarrow{f} Y$$

such that if W is any space, then $[W,(\beta)]$ is a long exact sequence. Under certain circumstances, we shall find that $[(\beta),W]$ and $[W,(\alpha)]$ are exact. We shall investigate this in § 1.2.

<u>Note</u>: Care should be taken to observe that $[X,Y]$ is a pointed set and not, in general, a group unless $X = SX'$ or $Y = \Omega Y'$. A sequence $A \xrightarrow{f} B \xrightarrow{g} C$ of pointed sets is exact if $f(A) = g^{-1}(*)$. In particular, g is a monomorphism if $f(A) = *$; but monomorphism means only that $g^{-1}(*) = *$. It does not mean (unless g is a group homomorphism) that g is $1 - 1$. Epimorphism does mean onto, however; and f is an epimorphism if and only if $g(B) = *$.

<u>Definition</u>: $p{:}E \to B$ is a fibre map if and only if p satisfies the covering homotopy property: if W is any space and

$$\begin{array}{ccc} W \times O & \xrightarrow{f} & E \\ \big\downarrow & & \big\downarrow p \\ W \times I & \xrightarrow{H} & B \end{array}$$

is a commutative diagram, then there exists $G{:}W \times I \to E$ making the diagram commute.

If p is a fibre map and $F = p^{-1}(*)$ then we call $F \xrightarrow{i} E \xrightarrow{p} B$ a fibration. We shall also call $F' \xrightarrow{i'} E' \xrightarrow{p'} B'$ a fibration if there exist homotopy equivalences $F \simeq F'$, $E \simeq E'$, $B \simeq B'$ making the total diagram homotopy commute.

<u>Theorem 1.1</u>: If $f: X \to Y$ is any map then $E_f \xrightarrow{j_f} X \xrightarrow{f} Y$ is a fibration.

<u>Note</u>: This says that in (β) above any two consecutive maps yield a fibration.

<u>Proof</u>: Let $\tilde{E}_f = \{(x,\omega) \in X \times Y^I \mid f(x) = \omega(1)\}$. Let $\tilde{f}: \tilde{E}_f \to Y$ be given by $\tilde{f}(x,\omega) = \omega(0)$. We have maps $X \leftrightarrows \tilde{E}_f$ where

$$x \longmapsto (x,\omega_x) \qquad (\text{with } \omega_x(t) = f(x))$$

$$x \longleftarrow (x,\omega)\ .$$

Clearly $\circlearrowleft$ is identity and $\circlearrowright$ takes $(x,\omega) \to (x,\omega_x)$. Let $H_t(x,\omega) = (x,\omega^t)$ where $\omega^t(1-s) = \omega(1-st)$ so that $\omega^1 = \omega$, $\omega^0 = \omega_x$.

Then $X \simeq \tilde{E}_f$ since $H_o = \circlearrowright$ and $H_1 = 1_{\tilde{E}_f}$. Also

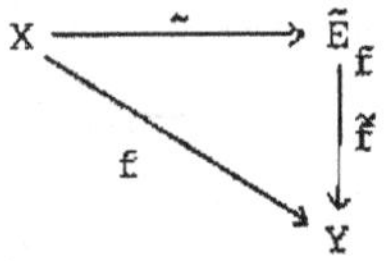

homotopy commutes.

Observe that $\tilde{f}^{-1}(*) = E_f$.

Finally we need to show that $\tilde{f}$ is a fibre map. Let $p:\tilde{E}_f \to X$, $q:\tilde{E}_f \to Y^I$ be the projections. Then given

$$\begin{array}{ccc} W \times 0 & \xrightarrow{\ h\ } & \tilde{E}_f \\ \big\downarrow & & \big\downarrow \tilde{f} \\ W \times I & \xrightarrow{\ H\ } & Y \end{array}$$

we define $G:W \times I \to \tilde{E}_f$ by $G(\omega,t) = (ph(\omega), \lambda_{\omega,t})$ where

$$\lambda_{\omega,t}(s) = \begin{cases} H(\omega, t - 2s) & 0 \le s \le t/2 \\ qh(\omega)\left(\frac{2s - t}{2 - t}\right) & t/2 \le s \le 1 \end{cases}$$

This is continuous since $H(\omega,0) = \tilde{f}h(\omega) = qh(\omega)(0)$ so $\lambda_{\omega,t}(t/2)$ is well-defined. Since $\lambda_{\omega,t}(1) = qh(\omega)(1) = fph(\omega), G(\omega,t) \in \tilde{E}_f$.

Also $G(\omega,0) = qh(\omega) = \tilde{f}h(\omega)$ and $\tilde{f}G(\omega,t) = \lambda_{\omega,t}(0) = H(\omega,t)$.

So the diagram will commute and the theorem is proved. (Except for the possibility of G being discontinuous. We simply remark that it is continuous for spaces we are interested in and we leave the exact conditions to point-set topologists.)

Next observe that the exactness of the sequence $\pi_n(\beta)$

β) $$\cdots \to \Omega Y \to E_f \to X \xrightarrow{f} Y$$

yields the following exact sequence

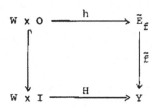

$$\begin{array}{ccccccccccc}
\pi_n(\Omega E_f) & \to & \pi_n(\Omega X) & \to & \pi_n(\Omega Y) & \to & \pi_n(E_f) & \to & \pi_n(X) & \to & \pi_n(Y) \\
\| \wr & & \| \wr & & \| \wr & \nearrow & & & & & \\
\pi_{n+1}(E_f) & \to & \pi_{n+1}(X) & \to & \pi_{n+1}(Y) & & & & & &
\end{array}$$

(We use the fact that $\pi_{n+1}(X) = [S^{n+1}, X] = [S(S^n), X] = [S^n, \Omega X] = \pi_n(\Omega X)$.)

1.3 Some Consequences of the Serre Spectral Sequence

Under certain circumstances, a map which is not a homotopy equivalence looks like one in low dimensions. To make this more precise, observe:

Theorem 1.2: If $f:A \to B$ then $f_*:H_i(A) \to H_i(B)$ is an isomorphism for $i < n$ and an epimorphism for $i \leq n$ if and only if the same is true of $f_*:\pi_i(A) \to \pi_i(B)$.

Proof: Consider the inclusion of A into the mapping cylinder $j:A \to Z_f$. Then

$$H_i(f) \text{ is an } \begin{cases} \text{iso} & i < n \\ \text{epi} & i \leq n \end{cases} \Leftrightarrow H_i(j) \text{ is same } \Leftrightarrow H_i(Z_f,A) = 0 \text{ for } i \leq n$$

$$\Updownarrow \text{ (Theorem 0.2)}$$

$$\pi_i(f) \text{ is same } \Leftrightarrow \pi_i(j) \text{ is an } \begin{cases} \text{iso } i < n \\ \text{epi } i \leq n \end{cases} \Leftrightarrow \pi_i(Z_f,A) = 0 \text{ for } i \leq n.$$

Corollary 1.3: The cofibre of f is n-connected if and only if the fibre of f is (n-1)-connected.

Definition: We say that f is n-connected in this case.

Remark: In all the above A and B are 1-connected. For A,B not 1-connected, there are examples where $C_f \simeq *$ but E_f is not 2-connected: as a group $\pi_2(S^1 \vee S^2) \simeq Z[t,t^{-1}]$ (a polynomial algebra on one variable and its inverse). Let $\theta:S^2 \to S^1 \vee S^2$ represent $2t - 1$. Then $H_2(\theta)$ is an isomorphism so $X = C_\theta$ has the same homology as S^1 . Then there is a map $f:S^1 \to X$ which is a homology isomorphism and in fact $C_f/S^1 \simeq *$, but $\pi_2(f)$ is <u>not</u> an isomorphism and E_f is not 2-connected.

We recall the Serre Spectral Sequence (e.g. as outlined in Spanier):

Theorem 1.4 (Serre): Let $F \xrightarrow{i} E \xrightarrow{p} B$ be a fibration with B 1-connected. Then there is a spectral sequence $\{E^r, d^r\}$ with $E^2_{s,t} = H_s(B;H_t(F))$ converging to $H_*(E)$; and a spectral sequence $\{\tilde{E}^r, \tilde{d}^r\}$ with $\tilde{E}^2_{s,t} = H_s(B,*;H_t(F))$ converging to $H_*(E,F)$. The edge homomorphisms are those induced by p_* and i_* .

Corollary 1.5: If B is $(n-1)$-connected and F is $(m-1)$-connected then for $\bar{p}:(E,F) \to (B,*), H_i(\bar{p})$ is an $\begin{cases} \text{iso} & \text{for } i < m+n \\ \text{epi} & \text{for } i \leq m+n \end{cases}$.

Proof: Picture the spectral sequence $\{\tilde{E}_r, \tilde{d}_r\}$ of $F \to E \to B$ converging to $H_*(E,F)$.

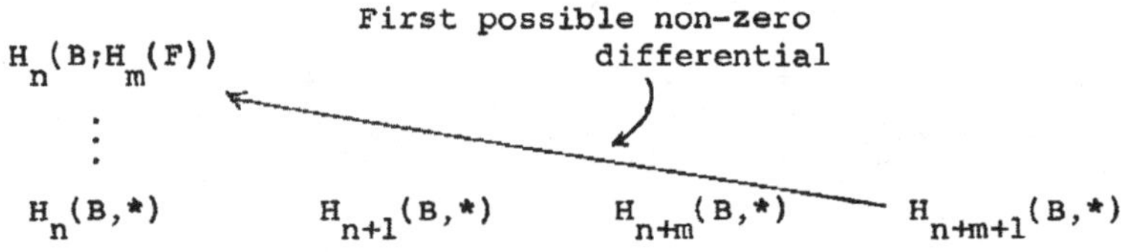

Thus since the edge homomorphism is $H_*(\bar{p})$ the result follows.

Notice that we cannot say that $\bar{p}:(E,F) \to (B,*)$ is $(m+n)$-connected. This does not make sense in the relative case: $\pi_i(\bar{p})$ is an isomorphism for all i although $H_i(\bar{p})$ is not. Conversely if $A \subset X$ satisfies HEP then $f:(X,A) \to (X/A,*)$ induces a homology isomorphism but not a homotopy isomorphism in all dimensions.

We can, however, put it this way:

$$\tilde{p}:E/F \to B \quad \text{induces an} \begin{cases} \text{iso} & \text{for } i < m+n \\ \text{epi} & \text{for } i \leq m+n \end{cases}$$

Thus $\tilde{p}:E/F \to B$ is $(m+n)$-connected.

We now wish to look at the dual problem: If $X \xrightarrow{f} Y \xrightarrow{i} C_f$ is a cofibration then there is an induced map $\rho:X \to E_i$. How close are X and E_i; i.e., how connected is ρ?

Look at Corollary 1.5 as follows:

For $i \leq n+m-1$, replace $H_i(E/F)$ by $H_i(B)$ in the exact homology sequence for $F \to E \to E/F$ yielding

$$H_i(F) \to H_i(E) \to H_i(B) \to H_{i-1}(F) \to \cdots$$

exact for $i \leq m+n-1$.

If $f:X \to Y$, let $E = E_{i_f} = \{(y,\omega) \in Y \times PC_f \mid \omega(1) = y\}$. Define $\rho:X \to E$ by $\rho(x) = (f(x),\omega_x)$ where $\omega_x(t) = (1-t,x)$ so that $\omega_x(0) = *, \omega_x(1) = f(x)$. Thus

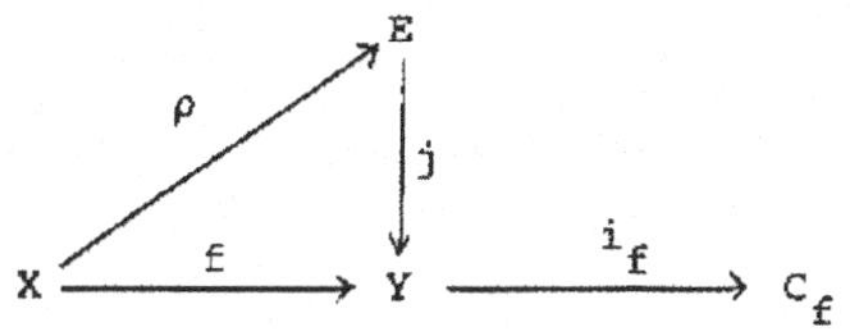

is commutative. Assume that X is $(n-1)$-connected and C_f is $(m-1)$-connected. From the exact homotopy sequence for $E \to Y \to C_f$ and homology for $X \to Y \to C_f$ and the Hurewicz theorem we find that E is also $(n-1)$-connected. But from the above we have both of the following sequences exact and the diagram commutative for

$k \leq n + m - 1$.

$$\begin{array}{ccccccccc}
H_k(Y) & \to & H_k(C_f) & \to & H_{k-1}(E) & \to & H_k(Y) & \to & \cdots \\
\| & & \| & & \uparrow \rho_* & & \| & & \\
H_k(Y) & \to & H_k(C_f) & \to & H_{k-1}(X) & \to & H_k(Y) & \to & \cdots .
\end{array}$$

Thus $H_k(\rho)$ is an isomorphism for $k \leq n + m - 2$ (but not necessarily an epimorphism for $k = n + m - 1$). Thus

Theorem 1.6: Given $X \xrightarrow{f} Y \xrightarrow{i} C_f$ with X $(n-1)$-connected and C_f $(m-1)$-connected then the induced map $\rho : X \to E_i$ is $(n+m-2)$-connected. Hence there is an exact homotopy sequence

$$\pi_{n+m-3}(X) \to \pi_{n+m-3}(Y) \to \pi_{n+m-3}(C_f) \to \pi_{n+m-4}(X) \to \cdots .$$

1.4 Getting to the Stable Range

We shall now proceed to make great use of Theorem 1.6. This will be the essential tool in getting to a stable situation. The idea is roughly that if a space is n-connected then its properties up to dimension $2n - \varepsilon$ ($\varepsilon = 0, 1,$ or 2 usually) are stable: e.g. suspending gives an isomorphism $H_i \to H_{i+1}$ and $\pi_i \to \pi_{i+1}$; looping gives an isomorphism $\pi_i \to \pi_{i-1}$ and $H_i \to H_{i-1}$. Within this range, fibrations and cofibrations "look the same." These ideas will become more precise in this section. From now on, the statement $X \subset Y$ will assume that (Y,X) has the HEP.

Theorem 1.7 (Blakers-Massey): If $X \subset Y$ and X is $(n-1)$-connected and Y/X is $(m-1)$-connected then

$$\varphi: \pi_i(Y,X) \to \pi_i(Y/X) \quad \text{is an} \quad \begin{cases} \text{iso} & \text{for } i < m+n-1 \\ \text{epi} & \text{for } i \le m+n-1 . \end{cases}$$

Proof: We have

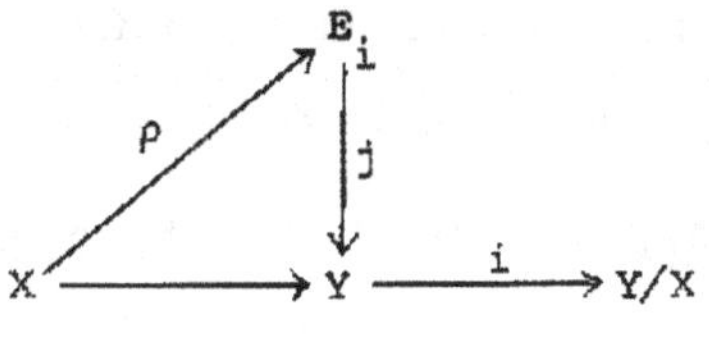

where ρ is $(n+m-2)$-connected. The following is an exact diagram (rows are exact and the diagram commutes):

$$\begin{array}{ccccccccc}
\pi_i(E_j) & \to & \pi_i(Y) & \to & \pi_i(Y/X) & \to & \pi_{i-1}(E_j) & \to & \pi_{i-1}(Y) \\
\rho_* \uparrow & & \| & & \varphi \uparrow & & \rho'_* \uparrow & & \| \\
\pi_i(X) & \to & \pi_i(Y) & \to & \pi_i(Y,X) & \to & \pi_{i-1}(X) & \to & \pi_{i-1}(Y)
\end{array}$$

We apply the 5-lemma:

for $i \leq n + m - 2$ ρ_*' is $\cong$ and ρ_* is epi so φ is $\cong$;

for $i = n + m - 1$ ρ_*' is epi so φ is epi .

Theorem 1.8: If A is (n-1)-connected and B is (m-1)-connected and $n \leq m$ then $\pi_i(A \vee B) \cong \pi_i(A) \oplus \pi_i(B) \oplus \pi_{i+1}(A \wedge B)$ for $i \leq m + 2n - 3$.

Proof: In the sequence $\pi_i(A \vee B) \to \pi_i(A \times B) \to \pi_i(A \times B, A \vee B) \to \ldots$ we observe that $\pi_i(A \times B) \cong \pi_i(A) \oplus \pi_i(B)$, and $\pi_i(A \vee B)$ contains this as a direct summand because A and B are retracts of $A \vee B$. Thus we have split exact sequences:

$$0 \to \pi_{i+1}(A \times B, A \vee B) \to \pi_i(A \vee B) \leftrightarrows \pi_i(A) \oplus \pi_i(B) \to 0$$

so

$$\pi_i(A \vee B) \cong \pi_i(A) \oplus \pi_i(B) \oplus \pi_{i+1}(A \times B, A \vee B) .$$

But $A \vee B$ is (n-1)-connected and using the Künneth formula we observe that $A \times B/A \vee B = A \wedge B$ is (m+n-1)-connected so $\pi_{i+1}(A \times B, A \vee B) \cong \pi_{i+1}(A \wedge B)$ for $i + 1 \leq m + 2n - 2$ applying Theorem 1.8.

Another place we use Theorem 1.6 is in the very important Freudenthal Suspension Theorem.

Theorem 1.9 (Freudenthal Suspension): If X is (n-1)-connected then

$$S:\pi_i(X) \to \pi_{i+1}(SX) \text{ is an } \begin{cases} \text{iso for } i < 2n-1 \\ \text{epi for } i \leq 2n-1 . \end{cases}$$

<u>Proof</u>: Look at the cofibration $X \xrightarrow{f} * \xrightarrow{i} SX$.

$$E_i = \Omega SX \quad \text{and} \quad \rho: X \to \Omega SX$$

is the map $\rho(x)(t) = t \wedge x$ the adjoint to the identity $SX \to SX$. Thus the composite $\pi_i(X) \xrightarrow{\rho_*} \pi_i(\Omega SX) = \pi_{i+1}(SX)$ is the suspension S . But ρ is (2n-1)-connected by Theorem 1.6 so

$$\rho_* \text{ is an } \begin{cases} \text{iso for } i < 2n-1 \\ \text{epi for } i \leq 2n-1 , \end{cases}$$

hence S is also.

We now introduce the type of space which will be most convenient for studying homotopy problems.

<u>Definition</u>: A CW complex is a space X together with a sequence of subspaces X^n such that

1) for some indexing set J_n, $X^n = X^{n-1} \bigcup_{\alpha \in J_n} e^n_\alpha$ where each e^n_α is an n-cell; i.e. there is an onto map $\varphi_\alpha : I^n \to e^n_\alpha$ which, restricted to interior I^n , is a homeomorphism onto $\overset{\circ}{e}{}^n_\alpha$. The boundary is $\dot{e}^n_\alpha = e^n_\alpha - \overset{\circ}{e}{}^n_\alpha$. Then $\overset{\circ}{e}{}^n_\alpha \cap (X^n \bigcup_{\beta \neq \alpha} e^n_\beta) = \emptyset$ and each $\dot{e}^n_\alpha$ is contained in a finite union of cells of dimension $< n$. Since $\dot{e}^n_\alpha$ is homeomorphic to S^{n-1} , there is defined a family of "characteristic maps" $\overset{\circ}{\varphi}_\alpha : S^{n-1} \to X^{n-1}$ and $C_{\overset{\circ}{\varphi}_{\alpha_0}} \cong X^{n-1} \cup e^n_{\alpha_0}$

2) X^0 is discrete

3) $X = \bigcup_{n=0}^{\infty} X^n$ and $O \subset X$ is open if and only if $O \cap X^n$ is open in X^n for all n .

Remarks:

i) 3) defines the "weak topology" on X with respect to the subspaces X^n . Observe that the topology remains the same if the word "open" is replaced by "closed."

ii) If X^{n-1} is connected, then $X^n/X^{n-1} \cong \bigvee_{J_n} S^n$ a wedge of n-spheres. Let $\varphi = \bigvee \dot{\varphi}_\alpha : \bigvee_{J_n} S^{n-1} \to X^{n-1}$. Then $C_\varphi \simeq X^n$. Thus there is a cofibration $\bigvee S^{n-1} \to X^{n-1} \to X^n$. We will find this particularly useful.

iii) A subcomplex A of a CW complex X , is a CW complex A such that $A^n \subset X^n$ and $A^n = A^{n-1} \bigcup_{\bar{J}_n} e^n_\alpha$ where $\bar{J}_n \subset J_n$. An extension of a remark in Chapter 0 yields the fact that (X,A) has the HEP.

Definition: We define the dimension of X by $\dim X \leq n$ if $X = X^n$.

1.5 CW Spaces

We call a space Y a CW space if and only if there is some CW complex $X \simeq Y$. We set $\dim Y = \min_{X \simeq Y} \dim X$. Since all theorems are statements only up to homotopy type any proof need involve only a CW complex X and the statement holds true for the CW space $Y \simeq X$. Observe that a CW complex may have a smaller dimension when considered as a CW space, but that will not matter because hypotheses have the form "dimension $\leq n$." For example, we have the following useful fact about CW spaces.

Lemma 1.10: If X is a connected CW space of dimension $\leq n$ and Y is an n-connected space, then $[X,Y] = 0$.

Proof: By induction on n. If $n = 1$, then $X \simeq \bigvee_\alpha S^1$ and $[X,Y] = [\bigvee_\alpha S^1,Y] = \prod_\alpha [S^1,Y] = \prod_\alpha \pi_1(Y) = 0$. Assume the lemma is true up to $(n-1)$. Let X and Y be as in the hypothesis. Then there is a map $f: \bigvee_\alpha S^{n-1} \to X^{n-1}$ with $C_f \simeq X$, so $X^{n-1} \to X \to \bigvee_\alpha S^n$ is a cofibration so $[X^{n-1},Y] \leftarrow [X,Y] \leftarrow [\bigvee_\alpha S^n,Y]$ is exact

$$\begin{array}{ccc} [X^{n-1},Y] \leftarrow [X,Y] \leftarrow & [\bigvee_\alpha S^n,Y] & \\ \| & \| & \\ 0 \text{ by induction,} & \prod_\alpha [S^n,Y] = \prod_\alpha \pi_n(Y) = 0 & \end{array}$$

since Y is n-connected, so $[X,Y] = 0$.

By more geometric means, we can extend Lemma 1.10 to the infinite dimensional case:

Lemma 1.11: Let Y be a space and X a CW complex with n cells only for those n such that $\pi_n(Y) = 0$. Then $[X,Y] = 0$.

Proof: Let $f:X \to Y$; we wish to construct a homotopy $H:X \times I \to Y$ with $H(x,0) = f(x)$, $H(x,1) = *$. X is obtained from $*$ by adjoining various n-cells. Well-order the procedure and construct H inductively. $H|*$ is trivial. Let X' be a subcomplex of X and assume we are given $H|X'$, let $X'' = X' \cup e^n$. Since (X'',X') has the HEP, $H|X'$ can be extended to $\tilde{H}:X'' \times I \to Y$ with $\tilde{H}|X' = H|X'$ and $\tilde{H}(x,0) = f(x)$. $\tilde{H}(x,1) = *$ for $x \in X'$. $\tilde{H}(x,1) = *$ for $x \in \dot{e}^n$ so $\tilde{H}_1:(e^n,\dot{e}^n) \to (Y,*)$ represents an element in $\pi_n(Y) = 0$, so there is a homotopy $G:(e^n,\dot{e}^n) \times I \to Y$ with $G_1 = *$ and $G_o = \tilde{H}_1$. Now define $H|\overset{\circ}{e}^n$ as follows: write $\overset{\circ}{e}^n = S^{n-1} \wedge [0,1)$ where $[0,1)$ has base point 0 . For $(x,t) \in \overset{\circ}{e}^n$ let

$$H((x,t),s) = \begin{cases} \tilde{H}((x,t),\frac{s}{t}) & 0 \le s \le t \text{ , } t > 0 \\ G((x,t),\frac{s-t}{1-t}) & t \le s \le 1 \text{ .} \end{cases}$$

This is continuous, and extends $H|X'$ since as $t \to 1$, $G((x,t),s) \to *$ uniformly for all s . Furthermore, $H((x,t),1) = G((x,t),1)) = *$. Thus $H|X''$ is defined. Inductively, then we have constructed a homotopy $H:f \sim *$.

We recall a few facts about relative homotopy groups. An element $\alpha \in \pi_n(X,A)$ is represented by a map $f:(E^n,S^{n-1}) \to (X,A)$ where E^n is some n-cell and S^{n-1} its boundary. $\alpha = 0$ if and only if $f \sim f'$ rel S^{n-1} where $f'(E^n) \subset A$. Also recall the long exact sequence

$$\cdots \to \pi_{n+1}(X,A) \to \pi_n(A) \to \pi_n(X) \to \pi_n(X,A) \to \cdots$$

All are elementary facts to be found in Hu [1] or Spanier.

We can use these facts to study some properties of CW complexes. We can think of a CW complex as being built up from the empty set by adding one cell at a time. (Use the axiom of choice to well-order the procedure.) If $\theta : S^{n-1} \to A$ then C_θ will often be written as $A \cup_\theta e^n$. Observe

<u>Lemma 1.12</u>: If $X = A \cup_\theta e^n$ and $\pi_n(Y,B) = 0$, then any map $f:(X,A) \to (Y,B)$ is homotopic rel. A to some f' where $f'(X) \subset B$.

<u>Proof</u>: Let $i:(E^n, S^{n-1}) \to (X,A)$ be the obvious map where $i|E^n - S^{n-1}$ is a homeomorphism onto $e^n - A$. Then $[f \circ i] \in \pi_n(Y,B)$ represents 0 so there is some $h':(E^n, S^{n-1}) \to (X,A)$ homotopic rel. S^{n-1} to $f \circ i$ and $h'(E^n) \subset B$. Then since $h'|S^{n-1} = f \circ i|S^{n-1}$ we can define $f':(X,A) \to (Y,B)$ by $f'|A = f|A$, $f'|e^n - A = h' \circ i^{-1}$ and $f' \sim f$ rel. A.

<u>Lemma 1.13</u>: Let X be a CW complex of dimension $N \leq \infty$ such that $X^n = \{*\}$. Assume $\pi_i(Y,B) = 0$ for $n-1 < i < N$. Then if $j:B \to Y$ is the inclusion then $j_*:[X,B] \to [X,Y]$ is injective. If $\pi_i(Y,B) = 0$ for $n < i < N+1$ then j_* is surjective. In particular if $\pi_i(Y,B) = 0$ for $n \leq i < N+1$, then j_* is an isomorphism.

<u>Proof</u>: Let E be the fibre of the inclusion $j:B \to Y$. Then the sequence $[X,E] \to [X,B] \xrightarrow{j_*} [X,Y]$ is exact. Comparing the long exact homotopy sequences, we observe that $\pi_i(E) \cong \pi_{i+1}(Y,B)$. Thus if $\pi_i(Y,B) = 0$ for $n-1 < i < N$ then $\pi_i(E) = 0$ for $n < i < N+1$. Since X has i-cells for $n < i < N+1$, Lemma 1.11 yields the fact that $[X,E] = 0$. Thus j_* is injective.

Let $f:X \to Y$ and let $X'' = X' \cup e^i$ be one stage in the construction of X.

Assume that $f|X' \sim f'|X'$ where $f'(X') \subset B$. By the HEP for (X'',X') (see Chap. 0) there is some $\tilde{f}'':X'' \to Y$ with $\tilde{f}'' \sim f$ and $\tilde{f}''|X' = f'$. Then applying Lemma 1.14 to $\tilde{f}''$ we get $f'':X'' \to Y$ such that $f'' \sim f'$ rel X' and $f''(X'') \subset B$. Then cell by cell, we construct $\bar{f}:X \to Y$ with $\bar{f}(X) \subset B$. Since at each stage the homotopy remained fixed there is a homotopy $\bar{f} \sim f$ defined on all of X. Now let $g:X \to B$ be defined by $g(x) = \bar{f}(x)$. Then $j \circ g = \bar{f} \sim f$. Thus j_* is surjective.

Finally we can prove a most important result on CW complexes, the Whitehead Theorem. We recall that a map $f:X \to Y$ is called a weak homotopy equivalence if $f_*:\pi_*(X) \to \pi_*(Y)$ is an isomorphism. A homotopy equivalence is clearly a weak homotopy equivalence. J. H. C. Whitehead has proved the converse on CW complexes:

Theorem 1.14: $f:X \to Y$ is a homotopy equivalence if and only if it is a weak homotopy equivalence. Furthermore if X and Y are 1-connected, then these conditions hold if and only if $f_*:H_*(X) \to H_*(Y)$ is an isomorphism.

Proof: Let $i:X \subset Z_f$, $j:Y \subset Z_f$, $r:Z_f \to Y$ be the usual maps with Z_f the mapping cylinder. If f is a w.h.e., then so is i; thus $\pi_*(Z_f,X) = 0$ so by Lemma 1.15, $[Z_f,X] \xrightarrow{i_*} [Z_f,Z_f]$ is surjective. Thus there is a map $\varphi:Z_f \to X$ with $i \circ \varphi \sim 1_{Z_f}$. Then $g = \varphi \circ j:Y \to X$ is such that $f \circ g = f \circ g \circ j \sim r \circ i \circ \varphi \circ j \sim r \circ j \sim 1_Y$. Thus f a w.h.e. implies there is some g with $f \circ g \sim 1_Y$. But then g is also a w.h.e. so there is some k with $g\circ k \sim 1_Y$. Then $k \sim (f\circ g)\circ k = f\circ(g\circ k) \sim f$ so $g\circ f \sim g\circ k \sim 1_X$. Thus g is a homotopy inverse to f.

The final statement on homology follows immediately from Theorem 1.2.

It is obvious that if X and Y are CW spaces and $f:X \to Y$ then C_f is a CW space. The following is not obvious and not easy and we shall not prove it here. It will be useful to keep this in mind as we continue, although we shall not use it.

<u>Theorem 1.15 (Milnor)</u>: If X and Y are CW spaces and $f:X \to Y$ then E_f is a CW space. In particular if $X = *$ then $E_f = \Omega Y$ is a CW space.

The proof for ΩY can be found in <u>Milnor</u>. The general case is unpublished.

<u>Theorem 1.16</u>: Let $f:X \to Y$ where X is $(n-1)$-connected and C_f is $(m-1)$-connected. If W is an r-dimensional CW space where $r \leqq n + m - 2$, then $[W,X] \xrightarrow{f_*} [W,Y] \xrightarrow{i_{f*}} [W,C_f]$ is exact. If Y is $(\ell-1)$-connected and $r \leqq n + \ell - 1$ also, there is a long exact sequence continuing to the right.

<u>Proof</u>: We have this diagram

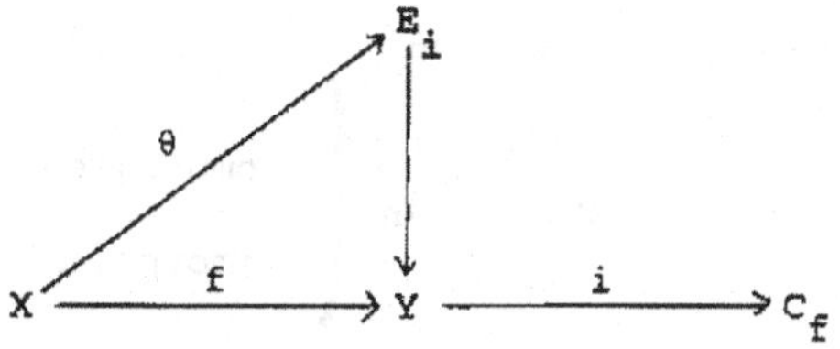

θ is (n+m-2)-connected by Theorem 1.6. Thus from Lemma 1.13, $\theta_*:[W,X] \to [W,E_i]$ is an epimorphism if $r \leqq n + m - 2$. But

$$[W,E_i] \to [W,Y] \to [W,C_f]$$

is always exact, hence if $r \leq n + m - 2$ then $[W,X] \to [W,Y] \to [W,C_f]$ is exact. Applying this to the cofibration $Y \to C_f \to SX$ completes the proof.

<u>Theorem 1.17</u>: Let W be a CW space of dimension r. Let $f:X \to Y$ be n-connected. Then

$$f_*:[W,X] \to [W,Y] \text{ is a } \begin{cases} \text{monomorphism if } r < n \\ \text{epimorphism if } r \leq n \end{cases}$$

<u>Proof</u>: $[W,E_f] \to [W,X] \xrightarrow{f_*} [W,Y]$ is exact. Since E_f is (n-1)-connected, $[W,E_f] = 0$ if $r < n$, hence f_* is a monomorphism. Since X is connected and C_f is n-connected

$[W,X] \xrightarrow{f_*} [W,Y] \xrightarrow{{}^i f_*} [W,C_f]$ is exact for $r \leq n$ by Theorem 1.16 and $[W,C_f] = 0$ by Lemma 1.11. Thus f_* is an epimorphism for $r \leq n$.

As a consequence we get theorems such as the following:

<u>Theorem 1.18</u>: If A is (n-1)-connected and B is (m-1)-connected, and X is a CW space of dimension r, then

$$[X,A \vee B] \to [X,A] \oplus [X,B] \text{ is an } \begin{cases} \text{isomorphism if } r < m + n - 1 \\ \text{epimorphism if } r \leq m + n - 1 \end{cases}$$

Proof: The map $[X, A \vee B] \to [X,A] \oplus [X,B] \cong [X, A \times B]$ is induced by $i: A \vee B \to A \times B$ which is $(m+n-1)$-connected since $C_i \simeq A \wedge B$.

We make one concession to point-set topology by proving

Lemma 1.19: Assume $Y = \bigcup_{i=1}^{\infty} Y_i$ has the weak topology. Assume $Y_i \subset Y_{i+1}$ and the Y_i are T_1-spaces. Then for X compact, $F(X,Y) = \varinjlim F(X,Y_n)$, the direct limit, and $[X,Y] = \varinjlim [X,Y_n]$.

Proof: Since $\varinjlim F(X,Y_n) \subset F(X,Y)$, for the first part it suffices to show that for any $f: X \to Y$ there is some n with $f(X) \subset Y_n$, if X is compact. Arguing by contradiction, assume $f(X) \not\subset Y_n$ for any n. Choose $y_n \in f(X) - Y_n$. Set $A = \{y_n\}_{n=1}^{\infty}$. For any $y \in Y$ $(A - \{y\}) \cap Y_n$ is finite, hence, since Y_n is T_1, closed. Thus $A - \{y\}$ is closed so y is not a limit point of A. So A is an infinite subset of $f(X)$ with no limit points. Thus $f(X)$ is non-compact. So X is non-compact. This contradicts the hypothesis.

We observe if X is compact then so is $X \times I$ so any homotopy takes place in some Y_n and the second part of the lemma follows.

From this lemma it immediately follows that $\pi_n(\bigcup_{i=1}^{\infty} Y_i) = \varinjlim \pi_n(Y_i)$ since S^n is compact.

We can extend Theorem 1.17 by induction to a finite wedge and then by Lemma 1.19 to a countable wedge.

Corollary 1.20: If $\{A_i\}_{i=1}^{\infty}$ is a countable collection of $(n-1)$-connected spaces and X is a compact CW space of dimension $\leq 2n - 2$ then

$$[X, \bigvee_{i=1}^{\infty} A_\alpha] = \sum_{i=1}^{\infty} [X,A_\alpha] .$$

<u>Proof</u>: $\bigvee_{i=1}^{\infty} A_i = \bigcup_{j=1}^{\infty} \bigvee_{i=1}^{j} A_i$ so for $m \leq 2n - 2$

$$\pi_m(\bigvee_{i=1}^{\infty} A_i) = \varinjlim_{j\to\infty} \pi_m(\bigvee_{i=1}^{j} A_i) = \varinjlim_{j\to\infty} \sum_{i=1}^{j} \pi_m(A_i) = \sum_{i=1}^{\infty} \pi_m(A_i) .$$

Then using the techniques of Corollary 1.10 extend to CW complexes of dimension $\leq 2n - 2$.

We observe that the theorem fails for X non-compact. For example if $X = \bigvee_1^{\infty} S^n$, then $[1_X] \in [X,X]$ but $1_X \notin \sum_1^{\infty} [X,S^n]$.

<u>Theorem 1.21 (Generalized Freudenthal)</u>: If X is (n-1)-connected and $\dim Y \leq r$ then $S:[Y,X] \to [SY,SX]$ is an

$$\begin{cases} \text{iso} & \text{if } r < 2n - 1 \\ \text{epi} & \text{if } r \leq 2n - 1 . \end{cases}$$

<u>Proof</u>: As we observed in the proof of Theorem 1.12, $\rho : X \to \Omega SX$ is (2n-1)-connected, where ρ is the adjoint of the identity $SX \to SX$. Since the composition

$$[Y,X] \xrightarrow{\rho_*} [Y,\Omega\, SX] \cong [SY,SX]$$

is the suspension, applying Theorem 1.19 to ρ finishes the proof.

Two "stability" theorems we will need later are the following:

Corollary 1.22: If X is $(n-1)$-connected and $\dim Y \leq r$, then the suspension map $S:F(Y,X) \to F(SY,SX)$ is $(2n-r-1)$-connected.

Proof: S induces

$$\begin{array}{ccc} S_*: \pi_i(F(Y,X)) & \to & \pi_i(F(SY,SX)) \\ \| \wr & & \| \wr \\ S: [S^iY,X] & \to & [S^{i+1}Y,SX] \end{array}$$

and thus is an isomorphism for $i + r < 2n - 1$ and an epimorphism for $i + r \leq 2n - 1$.

Notice that $f:X \to Y$ n-connected implies Sf is $(n+1)$-connected since $C_{Sf} \cong SC_f$ and Ωf is $(n-1)$-connected since $E_{\Omega f} \simeq \Omega E_f$.

Corollary 1.23: The map $X \to \Omega^r S^r X$ is $(2n-1)$-connected, if X is $(n-1)$-connected.

Proof: Taking $Y = S^0$ above yields $X \to \Omega SX$ $(2n-1)$-connected. Thus $SX \to \Omega S^2X$ is $(2n+1)$-connected so $\Omega SX \to \Omega^2 S^2 X$ is 2n-connected. Thus $X \to \Omega SX \to \Omega^2 S^2 X \to \ldots \to \Omega^r S^r X$ is $(2n-1)$-connected.

Stability, for us, will refer to those cases in which $[X,Y] \to [SX,SY]$ is an isomorphism. We can now put our previous results together to "stabilize" $[X,Y]$:

Corollary 1.24: If $\dim Y \leq n$ then $[S^jY,S^jX] \to [S^{j+1}Y,S^{j+1}X]$ is an isomorphism for $j \geq n + 2$ (regardless of the connectivity of X).

Definition: $\{Y,X\} = \varinjlim [S^jY,S^jX]$ is the set of S-maps from Y to X. Observe that if $\dim Y \leq n$, $\{Y,X\} = [S^jY,S^jX]$ for $j \geq n + 2$.

One of the most useful aspects of S-maps comes from the following theorem.

<u>Theorem 1.25</u>: If $X \to Y \to C_f$ is a cofibration, then $\{W,X\} \to \{W,Y\} \to \{W,C_f\}$ is exact, for any finite dimensional CW space W.

<u>Proof:</u> $[S^nW,S^nX] \to [S^nW,S^nY] \to [S^nW,S^nC_f]$ is exact if $\dim S^nW \leq 2n - 2$ since S^nX and S^nC_f are $(n-1)$-connected. Thus it is exact if $\dim W \leq n - 2$. So choose $n \geq \dim W + 2$ and the sequence is exact and yields (by the above) $\{W,X\} \to \{W,Y\} \to \{W, C_f\}$ exact.

<u>Definition</u>: $C^{\mathcal{S}}$ is the category whose objects are finite dimensional CW spaces and whose morphisms are S maps $\{\ ,\ \}$.

CHAPTER 2. EILENBERG - MacLANE SPACES AND SPECTRA

2.1 Construction of Certain Spaces

In this section we shall construct certain spaces having the property that their homotopy (Eilenberg - MacLane spaces) or homology (Moore spaces) groups vanish in every dimension except one. Based on these spaces, we will have a procedure for "dismantling" a given space to study its homology based on its homotopy, or vice versa. By taking fibrations or cofibrations with these spaces, we shall have a means of killing off homotopy or homology groups one at a time.

Definition: An Eilenberg - MacLane space of type (π,n) is a CW space $K(\pi,n)$ such that

$$\pi_i(K(\pi,n)) = \begin{cases} 0 & i \neq n \\ \pi & i = n \end{cases}$$

A Moore space of type (π,n) is a CW space $M(\pi,n)$ such that

$$\tilde{H}_i(M(\pi,n)) = \begin{cases} 0 & i \neq n \\ \pi & i = n \end{cases}$$ and $\pi_1(M(\pi,n))$ is abelian.

Clearly if $n \geq 2$, the existence of a $K(\pi,n)$ requires that π be abelian. The existence of an $M(\pi,n)$ always requires π to be abelian. These conditions are almost sufficient.

Theorem 2.1:

a) If π is abelian and $n \geq 2$, then $K(\pi,n)$'s and $M(\pi,n)$'s exist and are unique (up to homotopy type);

b) If π is any group, then $K(\pi,1)$ exists and is unique, and if π is abelian and $H^2(K(\pi,1)) = 0$ then $M(\pi,1)$'s exist (but are not necessarily unique);

c) $H^n(X;\pi)$ and $[X,K(\pi,n)]$ are naturally isomorphic for $n \geq 1$.

Since it will not be relevant to our work, we shall not prove b) but stick to the cases $n \geq 2$. (The case of $K(\pi,1)$ is of historical importance, however, as $H^*(K(\pi,1)) \cong H^*(\pi)$, the cohomology of the group π . Cf. MacLane.) For the proof of 2) regarding $M(\pi,1)$ se Varadarajan. Its non-uniqueness was shown in Chapter I where we saw two examples of $M(Z,1)$.

Observe that $\bigvee_{\alpha} S^n$ is a choice for $M(\sum_{\alpha} Z,n)$ (for $n \geq 2$). To construct an $M(G,n)$, $n \geq 2$, let

$$0 \to F \xrightarrow{f} H \to G \to 0$$

be a free abelian presentation of G . Let $F = \sum_{\beta \in B} Z$, $H = \sum_{\alpha \in A} Z$ and let f have the integral matrix form $((f_{\alpha\beta}))_{(\alpha,\beta) \in A \times B}$ for these bases. Then we have maps $f_{\alpha\beta}: S^n \to S^n$ (since the $f_{\alpha\beta}$ are integers). Let $\tilde{f} = ((f_{\alpha\beta})): \bigvee_{\beta \in B} S^n \to \bigvee_{\alpha \in A} S^n$.

The exact homology sequence for $\bigvee_{\beta \in B} S^n \xrightarrow{\tilde{f}} \bigvee_{\alpha \in A} S^n \to C_{\tilde{f}}$ yields the fact that $C_{\tilde{f}}$ is an $M(G,n)$; it follows from Theorem 1.16 that $C_{\tilde{f}}$ is 1-connected for $n \geq 2$.

For example, if $G = Z_q$ (integers modulo q), then $M(G,n) = S^n \cup_q e^{n+1}$, an n-sphere with an (n+1)-cell attached at its boundary (which is an n-sphere) by the map q .

As a corollary to the construction, we note that $M(G,n)$ may be taken to be a CW complex of dimension $\leq n + 1$.

We can construct $K(\pi,n)$ as follows: let $K_n = M(\pi,n)$. Then $\pi_i(K_n) = \begin{cases} 0 & i < n \\ \pi & i = n \end{cases}$. Inductively assume we have constructed $K_n \subset K_{n+1} \subset \ldots \subset K_m$ where

$$\pi_i(K_m) = \begin{cases} 0 & i < n \\ \pi & i = n \\ 0 & n < i \leq m \end{cases}$$

and $K_r \subset K_m$ induces a homotopy isomorphism up to degree r , for all $r \leq m$. Let $f_\alpha : S^{m+1} \to K_m$ be such that the $[f_\alpha]$ generate $\pi_{m+1}(K_m)$. Let $f = \bigvee f_\alpha : \bigvee S^{m+1} \to K_m$. Then let $K_{m+1} = C_f$. By Theorem 1.20 there is an exact sequence

$$\pi_i(\bigvee S^{m+1}) \xrightarrow{f_*} \pi_i(K_m) \xrightarrow{i_*} \pi_i(K_{m+1}) \to \pi_{i-1}(\bigvee S^{m+1}) \to \ldots$$

for $i \leq n + m - 1$. Since $\pi_i(\bigvee S^{m+1}) = 0$ for $i \leq m$ i_* is an isomorphism there. But $\pi_{m+1}(f_*)$ is onto and $\pi_m(\bigvee S^{m+1}) = 0$ so $\pi_{m+1}(K_{m+1}) = 0$. Thus we have constructed K_{m+1} inductively. Then with the weak topology $K = \bigcup_{m=n}^{\infty} K_m$ is a CW complex and using Lemma 1.11

$$\pi_i(K) = \varinjlim_m \pi_i(K_m) = \begin{cases} 0 & i \neq n \\ \pi & i = n \end{cases} .$$

Thus we have constructed a $K(\pi,n)$.

Next we shall show that for CW spaces X , $[X,K(\pi,n)] \simeq H^n(X;\pi)$. Now $H^n(K(\pi,n);\pi) \simeq \mathrm{Hom}(H_n(K(\pi,n)),\pi) \cong \mathrm{Hom}(\pi,\pi)$. Choose $\iota \in H^n(K(\pi,n);\pi)$ corresponding to 1_π . Define a natural transformation $T:[\ ,K(\pi,n)] \to H^n(\ ;\pi)$ by $T(X)[f] = f^*(\iota)$ for any $f:X \to K(\pi,n)$. By the choice of ι, $T(S^n)$ is an isomorphism. By the triviality of both sides $T(S^m)$ is an isomorphism for $m \neq n$. Since a product of isomorphisms is again an isomorphism, $T(\bigvee S^m)$ is for all m.

Let X be a CW space. Assume that $T(S^iX^r)$ has been shown to be an isomorphism for all i and for some fixed r. (E.g. we can start off with $r = 0$.) Then the Puppe sequence for the i-th suspension of the cofibration $\bigvee S^r \to X^r \to X^{r+1}$

$$\bigvee S^{r+i} \to S^iX^r \to S^iX^{r+1} \to \bigvee S^{r+i+1} \to S^{i+1}X^r ,$$

yields the exact diagram

$$\begin{array}{ccccccccc}
[\bigvee S^{r+i},K(\pi,n)] & \to & [S^iX^r,K(\pi,n)] & \to & [S^iX^{r+1},K(\pi,n)] & \to & [\bigvee S^{r+i+1},K(\pi,n)] & \to & [S^{i+1}X^r,K(\pi,n)] \\
\downarrow T(\bigvee S^{r+i}) & & \downarrow T(S^iX^r) & & \downarrow T(S^iX^{r+1}) & & \downarrow T(\bigvee S^{r+i+1}) & & \downarrow T(S^{i+1}X^r) \\
H^n(\bigvee S^{r+i};\pi) & \to & H^n(S^iX^r;\pi) & \to & H^n(S^iX^{r+1};\pi) & \to & H^n(\bigvee S^{r+i+1};\pi) & \to & H^n(S^{i+1}X^r;\pi)
\end{array}$$

Since the first, second, fourth and fifth are isomorphisms, so is the third, hence $T(X)$ is an isomorphism for finite dimensional CW spaces X.

Observe that the above did not require $K(\pi,n)$ to be a CW space. In particular, given some $K(\pi,n+1)$,

$$\pi_i(\Omega K(\pi,n+1)) \cong \pi_{i+1}(K(\pi,n+1)) = \begin{cases} \pi & i = n \\ 0 & i \neq n \end{cases}$$ so we have

$T':[\ ,\Omega K(\pi,n)] \to H^n(\ ;\pi)$ defined with $T'(X)$ an isomorphism for finite dimensional CW spaces.

Let X be an arbitrary CW complex. Then X/X^{n+1} is a CW complex with i-cells only for $i \geq n+2$ and $i = 0$. Thus from Lemma 1.11, $[X/X^{n+1}, K(\pi,n+1)] = 0$ and $[S(X/X^{n+1}),K(\pi,n+1)] = 0$.

Applying $[\ ,K(\pi,n+1)]$ to the Barratt-Puppe sequence of $\iota : X^{n+1} \hookrightarrow X$ yields the exact sequence

$$\begin{array}{ccccccc}
[X/X^{n+1},K(\pi,n+1)] & \longleftarrow & [SX^{n+1},K(\pi,n+1)] & \longleftarrow & [SX,K(\pi,n+1)] & \longleftarrow & [S(X/X^{n+1}),K(\pi,n+1)] \\
\| & & \wr\| & & \wr\| & & \| \\
0 & & [X^{n+1},\Omega K(\pi,n+1)] & \xleftarrow{i\#} & [X,\Omega K(\pi,n+1)] & & 0 \\
 & & \Big\downarrow {\scriptstyle T'(X^{n+1})} & & \Big\downarrow {\scriptstyle T'(X)} & & \\
 & & H^n(X^{n+1};\pi) & \xleftarrow{i^*} & H^n(X;\pi) & &
\end{array}$$

By the exactness, $i\#$ is an isomorphism. By the long exact cohomology sequence, i^* is an isomorphism. By the previous part $T'(X^{n+1})$ is an isomorphism, hence $T'(X):[X,\Omega K(\pi,n+1)] \to H^n(X;\pi)$ is an isomorphism.

In particular if $K'(\pi,n)$ is an Eilenberg-MacLane space of type (π,n), then $[K(\pi,n),\Omega K(\pi,n+1)] \cong H^n(K(\pi,n);\pi) \cong \mathrm{Hom}(\pi,\pi)$ choosing $f:K(\pi,n) \to \Omega K(\pi,n+1)$ representing $1_\pi : \pi \to \pi$, we see (details are in the more general Lemma 2.3) that f is a weak homotopy equivalence. We may as well assume f to be an inclusion. Then from Lemma 1.13, we see that for any CW space X, $[X,K(\pi,n)] \xrightarrow{f_*} [X,\Omega K(\pi,n+1)]$ is an isomorphism. Thus composing f_* with $T'(X)$ yields $T(X):[X,K(\pi,n)] \to H^n(X;\pi)$ an isomorphism.

<u>Note</u>: We could have skipped some of this by using the fact (Theorem 1.15 which was not proved) that $\Omega K(\pi,n+1)$ is a CW space, hence is a choice for $K(\pi,n)$.

Next we observe the following useful property of Moore spaces.

Lemma 2.2: If $\pi_i(X) = 0$ for $i < n \geq 2$, then there is a map $h: M(\pi_n(X), n) \to X$ which is a π_n (and hence H_n) isomorphism.

Proof: Look at the construction of the "standard" $M(\pi, n); \pi = \pi_n(X)$: if $0 \to F \xrightarrow{f} G \xrightarrow{p} \pi \to 0$ is exact where F and G are free abelian, then there is a map $\tilde{f}: \bigvee_\alpha S^n \to \bigvee_\beta S^n$ such that $\pi_n(\tilde{f})$ represents f. $M(\pi, n)$ is $C_{\tilde{f}}$. For each β, there is a map $g_\beta: S^n \to X$ representing one of the generators of π. Then the map $g = \bigvee g_\beta : \bigvee_\beta S^n \to X$ is defined and $\pi_n(g)$ represents p. Then $g\tilde{f} \sim *$ (since $pf = 0$). Thus g may be lifted to a map $h: M(\pi, n) \to X$ which clearly induces a π_n (and hence H_n) isomorphism.

In particular it follows from this and from the Whitehead Theorem (1.14) that $M(\pi, n)$'s are unique for $n \geq 2$.

As a dual statement to Lemma 2.2 we have

Lemma 2.3: If X is an $(n-1)$-connected CW space then there is a map $f: X \to K(\pi_n(X), n)$ which induces a π_n (and hence H_n) isomorphism.

Proof: Let $\pi = \pi_n(X) \cong H_n(X)$. Then

$$[X, K(\pi, n)] \cong H^n(X; \pi) \cong \operatorname{Hom}(H_n(X), \pi) \cong \operatorname{Hom}(\pi, \pi) .$$

Choose $f: X \to K(\pi, n)$ corresponding to 1_π.

Now let us pause a moment and consider what it means to have $\varphi \in H^n(Y; \pi)$ correspond to $g: Y \to K(\pi, n)$. In particular g induces $g_*: H_n(Y) \to H_n(K(\pi, n)) = \pi$ so that $g_* \in \operatorname{Hom}(H_n(Y), \pi)$. Then the epimorphism $H^n(Y; \pi) \to \operatorname{Hom}(H_n(Y), \pi)$ sends φ to g_*. Thus in our situation

$$\begin{array}{ccc} H_n(X) & \xrightarrow{f_*} & H_n(K(\pi,n)) \\ \wr\| & & \wr\| \\ \pi & \xrightarrow{1_\pi} & \pi \end{array}$$

is a commutative diagram, hence f_* is an isomorphism.

Now we can prove the uniqueness of the $K(\pi,n)$'s. If X is any Eilenberg-MacLane space of type (π,n) and $K(\pi,n)$ is the standard one, then by Lemma 2.3 there is a map $f:X \to K(\pi,n)$ which induces a π_n-isomorphism. By the triviality in other dimensions, f is a w.h.e., hence by the Whitehead Theorem a homotopy equivalence. This completes the proof of Theorem 2.1.

2.2 Properties of Eilenberg-MacLane and Moore Spaces

We next prove some useful properties of CW spaces. First we observe

Theorem 2.4: Let X be a 2-connected CW space such that $H^i(X) = 0$ for $i > n$. Then for some integer r, $S^rX \simeq$ a CW complex of dimension $\leq n + r$. Conversely if $\dim X \leq n$ then $H^i(X) = 0$ for $i > n$.

Proof: By induction on t where X is $(n-t)$-connected: For $t = 0$ or 1, X must be a wedge of spheres and thus is unique. Assume the theorem proved for $1 \leq t \leq (n-m)$ and let X be $(m-1)$-connected. Then by Lemma 2.2 there is a map $h: M(\pi_m(X), m) \to X$ which is an H_m-isomorphism. Thus $h^*: H^*(X) \to H^*(M(\pi_m(X), m))$ is an epimorphism (by the naturality of the universal coefficient theorem). Thus $i_h^*: H^*(C_h) \to H^*(X)$ is a monomorphism so $H^i(C_h) = 0$ for $i > n$. On the other hand C_h is m-connected. By the inductive hypothesis, then, there exists a CW complex C' of dimension $\leq (n+r)$ such that $C' \simeq S^rC_h$. Now $SX \simeq C_{\sigma_h}$, $\sigma_h: C_h \to M(\pi_m(X), m+1)$ so $S^{r+1}X$ is the cone of a map

$$C' \simeq S^rC_h \to M(\pi_m(X), m+r+1)$$

and this cone clearly is a CW complex of dimension $\leq \max(r+n+1, m+r+2) = r + n + 1$. Thus $S^{r+1}X \simeq$ a CW complex of dimension $\leq n + r + 1$.

The converse is gotten by an easy induction argument.

(A much more geometric argument shows that r may be taken equal to 0 .)

The following is an immediate consequence of this and Lemma 1.14.

__Theorem 2.5__: If $H^i(Y) = 0$ for $i > n$ and $\pi_i(X) = 0$ for $i \leq n$ and Y is a CW space , then for some r $[S^rY, S^rX] = 0$. Thus $\{Y,X\} = 0$.

Notice that the hypotheses of the above theorem imply that $H^i(Y;\pi_i(X)) = 0$. We may wonder if that is not the most important point. In fact with much weaker hypotheses we can get a much sharper theorem by approaching the problem from the dual point of view: by fibrations and Eilenberg-MacLane spaces:

__Theorem 2.6__: Let X and Y be CW spaces with $H^i(Y;\pi_i(X)) = 0$ for all i . If either $\pi_i(X) = 0$ for sufficiently large i or Y is finite dimensional, then $[Y,X] = 0$.

__Proof__: We can prove this by induction: let $E_o = X$. Given E_n an n-connected space with $\pi_i(E_n) \cong \pi_i(X)$ for $i > n$, let $f:E_n \to K(\pi_{n+1}(X),n+1)$ induce $\pi_{n+1}(f)$, an isomorphism (using Lemma 2.3). Let E_{n+1} be the fibre of f . Then E_{n+1} is (n+1)-connected and $\pi_i(E_{n+1}) \cong \pi_i(E_n) \cong \pi_i(X)$ for $i > n + 1$.

Now if $\pi_i(X) = 0$ for $i \geq r$ then since Y is a CW space and $\pi_*(E_r) = 0$ we have $[Y,E_r] = 0$. If on the other hand Y is r-dimensional, then $[Y,E_r] = 0$.

Now we work backwards inductively. From the fibration $E_{n+1} \to E_n \to K(\pi_{n+1}(X),n+1)$, there is an exact sequence

$$[Y,E_{n+1}] \to [Y,E_n] \to [Y,K(\pi_{n+1}(X),n+1)] \ .$$

But $[Y,K(\pi_{n+1}(X),n+1)] = H^{n+1}(Y;\pi_{n+1}(X)) = 0$ so that $[Y,E_{n+1}] = 0$ implies that $[Y,E_n] = 0$. Thus inductively $[Y,X] = [Y,E_o] = 0$.

We have already observed that $\Omega K(\pi,n+1) \simeq K(\pi,n)$. This is very useful. We wish to consider more general cases of sequences of spaces A_n with $A_n \simeq \Omega A_{n+1}$. We generalize this to the idea of spectra.

<u>Definition</u>: A spectrum $\underline{X}$ is a sequence of spaces X_n and maps $\epsilon_n : SX_n \to X_{n+1}$ or, equivalently, $\tilde{\epsilon}_n : X_n \to \Omega X_{n+1}$. Examples:

1) $X_n = K(\pi,n), \tilde{\epsilon}_n : K(\pi,n) \simeq \Omega K(\pi,n+1)$ this gives $\underline{K}(\pi)$
2) W a space, $X_n = S^n W$, $\epsilon_n : S(S^n W) \simeq S^{n+1} W$ this gives $\underline{SW}$
3) $\underline{X}$ a spectrum $(\underline{X} \wedge W)_n = X_n \wedge W$
4) $\underline{X}$ a spectrum $(\underline{X}^d)_n = \begin{cases} X_{n+d} & n+d \geq 0 \\ * & n+d < 0 \end{cases}$

   $\underline{X}^d$ is the d-th suspension of $\underline{X}$ and is defined for <u>all</u> integers d .
5) if W is a compact space and $\underline{X}$ is a spectrum we can form the spectrum $F(W,\underline{X})$ whose n-th space is $F(W,X_n)$ (with the compact open topology). The maps are given by $SF(W,X_n) \xrightarrow{\alpha} F(W,SX_n) \xrightarrow{F(W,\epsilon_n)} F(W,X_{n+1})$ where $\alpha(t \wedge f)(w) = t \wedge f(w)$.

From now on we will assume that all spectra are CW spectra i.e., each space is a CW space. We can now define some functors on spectra:

Definition: $H_r(\underline{X}) = \varinjlim_n H_{r+n}(X_n)$

$$\pi_r(\underline{X}) = \varinjlim_n \pi_{r+n}(X_n)$$

$$H^r(\underline{X}) = \varprojlim_n H^{r+n}(X_n)$$

where the direct and inverse systems are given by

$$H_{r+n}(X_n) \cong H_{r+n+1}(SX_n) \xrightarrow{\epsilon_{n*}} H_{r+n+1}(X_{n+1})$$

$$\pi_{n+n}(X_n) \to \pi_{r+n+1}(SX_n) \xrightarrow{\epsilon_{n*}} \pi_{r+n+1}(X_{n+1})$$

$$H^{r+n+1}(X_{n+1}) \xrightarrow{\epsilon_n^*} H^{r+n+1}(SX_n) \cong H^{r+n}(X_n)$$

A map of spectra $f:\underline{X} \to \underline{Y}$ of degree r is a sequence of maps $f_n:X_n \to Y_{n+r}$, for n sufficiently large, such that the following diagram commutes

$$\begin{array}{ccc} SX_n & \xrightarrow{Sf_n} & SY_{n+r} \\ \downarrow{\scriptstyle \epsilon_n} & & \downarrow{\scriptstyle \epsilon_{n+r}} \\ X_{n+1} & \xrightarrow{f_{n+1}} & Y_{n+r+1} \end{array}$$

Then we can define the spectrum $\underline{C}_f$, the cone of f, with $(\underline{C}_f)_{n+r} = C_{f_n}$ with the map $SC_{f_n} = C_{Sf_n} \to C_{f_{n+1}}$ induced by the commutativity of the above diagram. (We can similarly define a spectrum $\underline{E}_f$, the fibre of f.)

The set of homotopy classes of maps of spectra of degree r will be denoted by $[\underline{X},\underline{Y}]^r$ or $[\underline{X},\underline{Y}]_{-r}$. We shall let $\{W,\underline{Y}\} = \varinjlim [S^nW,Y_n] = [\underline{S}W,\underline{Y}]$.

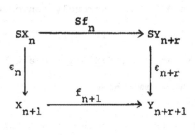

We shall find certain types of spectra as manageable as spaces.

Definition:

1) A spectrum $\underline{X}$ is convergent if and only if for some N, $\pi_i(\underline{X}) = 0$ for all $i \leq N$. ($\underline{X}$ will be called N-connected.)

2) A spectrum $\underline{X}$ is strongly convergent if and only if for some N each X_n is $(n+N)$-connected for n sufficiently large (hence $\underline{X}$ is N-connected), and furthermore, for all q the map $H_{q+k+1}(SX_k) \to H_{q+k+1}(X_{k+1})$ is an isomorphism for almost all k. (This last statement says that for all q, ε_k is $(q+k)$-connected for almost all k.)

3) $f:\underline{X} \to \underline{Y}$ is a weak homotopy equivalence (w.h.e.) if and only if it is of degree 0 and $f_*:\{W,\underline{X}\} \to \{W,\underline{Y}\}$ is an isomorphism for every finite CW complex W.

Theorem 2.7: If $\underline{X}$ and $\underline{Y}$ are strongly convergent spectra, then $f:\underline{X} \to \underline{Y}$ is a w.h.e. if and only if $f_*:H_*(\underline{X}) \to H_*(\underline{Y})$ is an isomorphism. For any $\underline{X}$ and $\underline{Y}$, $f:\underline{X} \to \underline{Y}$ is a w.h.e. if and only if $f_*:\pi_*(X) \to \pi_*(\underline{Y})$ is an isomorphism.

Proof: Let m be an integer. Choose an integer N such that

1) X_k and Y_k are 1-connected for $k \geq N$

2) for all $j \leq m$ the natural maps

$$\left.\begin{array}{l} H_{j+k}(X_k) \to H_j(\underline{X}) \\ H_{j+h}(Y_k) \to H_j(\underline{Y}) \end{array}\right\} \text{ are isomorphisms for all } k \geq N$$

and 3) For every finite CW complex of dimension $\leq m$ the maps

$$\left.\begin{array}{l} [S^kP,X_k] \to \{P,\underline{X}\} \\ [S^kP,Y_k] \to \{P,\underline{Y}\} \end{array}\right\} \text{ are isomorphisms for all } k \geq N .$$

The diagrams

$$\begin{array}{ccc} H_{j+k}(X_k) & \longrightarrow & H_j(\underline{X}) \\ \downarrow f_{k*} & & \downarrow f_* \\ H_{j+k}(Y_k) & \longrightarrow & H_j(\underline{Y}) \end{array}$$

$$\begin{array}{ccc} [S^kP, X_k] & \longrightarrow & \{P, \underline{X}\} \\ \downarrow f_{k*} & & \downarrow f_* \\ [S^kP, Y_k] & \longrightarrow & \{P, \underline{Y}\} \end{array}$$

commute.

If $k \geq N$, $j \leq m$ and $\dim P \leq m$, then the horizontal arrows are isomorphisms.

Now if f is a w.h.e. then $f_{k*}:[S^k(S^r), X_k] \to [S^k(S^r), Y_k]$ is an isomorphism for $r \leq m$ so $f_{k*}:\pi_{i+k}(X_k) \to \pi_{i+k}(Y_k)$ is an isomorphism for $i \leq m$ so $f_{k*}:H_{i+k}(X_k) \to H_{i+k}(Y_k)$ is an isomorphism for $i < m$ so $f_*:H_i(\underline{X}) \to H_i(\underline{Y})$ is an isomorphism for $i < m$. But m was arbitrary so f_* is an isomorphism for all i. We can work backwards to prove the converse once we prove the second statement.

If $f_*:\pi_*(\underline{X}) \to \pi_*(\underline{Y})$ is an isomorphism, then $f_*:\{S^n, \underline{X}\} \to \{S^n, \underline{Y}\}$ is an isomorphism for every sphere S^n. The result then follows by induction on the number of cells and by the 5-lemma in the obvious way.

Two useful types of spectra that we have seen before are the following.

Definition: A spectrum $\underline{X}$ is an S-spectrum iff $SX_n \simeq X_{n+1}$ for almost all n. A spectrum $\underline{X}$ is an Ω-spectrum iff $X_n \simeq \Omega X_{n+1}$ for almost all n.

<u>Theorem 2.8</u>:

a) S-spectra are strongly convergent.

b) Convergent Ω-spectra are strongly convergent.

c) Given a spectrum $\underline{X}$ there exists an Ω-spectrum X' w.h.e. to $\underline{X}$.

b) + c) $\Rightarrow$ d) If $\underline{X}$ is convergent then there exists a strongly convergent Ω-spectrum $\underline{X}'$ w.h.e. to $\underline{X}$.

<u>Proof</u>:

a) is trivial.

b) Assume $\pi_i(\underline{X}) = 0$ for $i \leq N$. Then assume $X_n \simeq \Omega X_{n+1}$ for all $n \geq M$. Then $\pi_{i+n}(X_n) = 0$ $i \leq N$ for all $n \geq M$. Thus each X_n is $(n+N)$-connected. Then we have $SX_n \simeq S\Omega X_{n+1} \xrightarrow{\epsilon_n} X_{n+1}$. But

$$\begin{array}{ccc} \pi_{i+n}(X_n) & \xrightarrow{S} & \pi_{i+n+1}(SX_n) \\ \wr\!\wr & & \downarrow \epsilon_{n*} \\ \pi_{i+n}(\Omega X_{n+1}) & \xrightarrow{\sim} & \pi_{i+n+1}(X_{n+1}) \end{array}$$

commutes. By Theorem 1.13 S is an

$$\begin{cases} \text{isomorphism for } i+n < 2(n+N+1) - 1 \\ \text{epimorphism for } i+n \leq 2(n+N+1) - 1 \end{cases}$$

Thus ϵ_{n*} is an

$$\begin{cases} \text{iso for } i < n+2N+1 \\ \text{epi for } i \leq n+2n+1 \end{cases}$$

. Thus ϵ_n is $n+(n+2N+1)$-connected. So for all $n \geq q - 2N - 1$, ϵ_n is $(n+q)$-connected. Thus $\underline{X}$ is strongly convergent.

c) Let $X_n' = \lim\limits_{\overrightarrow{r}} \Omega^r X_{n+r}$ where $\Omega^r X_{n+r} \xrightarrow{\Omega^r \epsilon_{n+r}} \Omega^r(\Omega X_{n+r+1}) = \Omega^{r+1} X_{n+r+1}$ gives the direct system.

Since $\Omega = F(S^1, \ )$ and S^1 is compact Lemma 1.19 yields $\Omega X'_{n+1} = \Omega \varinjlim_r \Omega^r X_{n+r+1} = \varinjlim_r \Omega^{r+1} X_{n+r+1} = \varinjlim_r \Omega^r X_{n+r} = X'_n$. Thus $\underline{X}'$ is an Ω-spectrum.

Define $f:\underline{X} \to \underline{X}'$ by $X_n = \Omega^0 X_n \to \varinjlim_r \Omega^r X_{n+r}$. Given n choose s so that $n + s > 0$. Then $\pi_n(\underline{X}) = \varinjlim_r \pi_{n+r}(X_r) =$ $\varinjlim_r \pi_{n+s}(\Omega^{r-s} X_r) \simeq \pi_{n+s}(\varinjlim_r \Omega^{r-s} X_r) = \pi_{n+s}(\varinjlim_r \Omega^r X_{r+s}) = \pi_{n+s}(X'_s) =$ $\pi_n(\underline{X}')$ and this isomorphism is that given by f_* . Using Theorem 2.7, f is a w.h.e.

Corollary 2.9: $\underline{K}(G)$ and $\underline{S}$ are strongly convergent.

At this point we note that not all Ω spectra are convergent. For example, Bott periodicity (Bott) says that $\Omega^2 U \simeq U$ and $\Omega^8 O \simeq O$ for the unitary and orthogonal groups. Thus there is a spectrum $\underline{X}$ with $X_{2n} = \Omega U$, $X_{2n+1} = U$ whence $\pi_{2m}(\underline{X}) \simeq Z$ for every integer m . (Similarly we make a spectrum out of O with $\pi_{4m}(\underline{X}) \simeq Z$ for every integer m .)

2.3 Generalized Homology Theories

Let C^2 be the category of pairs of spaces (X,A) satisfying HEP. Let C^* be the category of based spaces. In both cases the morphisms are homotopy classes of maps. We write $(X,\emptyset)$ as X. Let $\sigma: C^2 \to C^2$ be the functor given by $\sigma(X,A) = A$. Let $\mathfrak{G}$ be the category of abelian groups.

<u>Definition</u>: A generalized homology theory $\mathfrak{H}$ on C^2 is a sequence of functors $H_n: C^2 \to \mathfrak{G}$ and natural transformations $\partial_n: H_n \to H_{n-1}\circ\sigma$ such that

1) $\cdots \xrightarrow{\partial_{n+1}} H_n(A) \xrightarrow{i_*} H_n(X) \xrightarrow{j_*} H_n(X,A) \xrightarrow{\partial_n} H_{n-1}(A) \to \cdots$ is exact for every pair (X,A) and

2) $H_n(X,A) \xrightarrow{p_*} H_n(X/A,*)$ is an isomorphism, p the projection.

<u>Note</u>: By convention $X/\emptyset = X\cup *$, a disjoint union with $*$ acting as a basepoint.

<u>Definition</u>: A reduced generalized homology theory $\tilde{\mathfrak{H}}$ on C^* is a sequence of functor $\tilde{H}_n: C^* \to \mathfrak{G}$ and natural transformations $\sigma_n: H_n \to H_{n+1}\circ S$ such that

1) if $A \subset X$ satisfies HEP then $\tilde{H}_n(A) \to \tilde{H}_n(X) \to \tilde{H}_n(X/A)$ is exact. And

2) $\sigma_n(X)$ is an isomorphism for every object X of C^*.

The coefficients of a theory are $H_*(*)$ or $\tilde{H}_*(S^0)$. Observe that if X is a non-empty space then $x_0 \in X$ is a retract of X hence $H_*(x_0) \cong H_*(*)$ is a summand of $H_*(X)$ so $H_*(X) \cong H_*(x_0) \oplus H_*(X,x_0)$.

<u>Claim</u>: For $X \in C^*$ define $\tilde{H}_n(X) = H_n(X,*)$. Then this gives us a reduced homology theory: since $\tilde{H}_n(X/A) = H_n(X/A,*) \cong H_n(X,A)$ we have the exact sequence $H_n(A) \to H_n(X) \to \tilde{H}_n(X/A) \to H_{n-1}(A) \to H_{n-1}(X)$, but

$$\begin{array}{ccccc} H_n(A) & \cong & H_n(*) & \oplus & \tilde{H}_n(A) \\ \downarrow & & \wr\!\downarrow & & \downarrow \\ H_n(X) & \cong & H_n(*) & \oplus & \tilde{H}_n(X) \end{array}$$

commutes so $\tilde{H}_n(A) \to \tilde{H}_n(X) \to \tilde{H}_n(X/A) \to \tilde{H}_{n-1}(A) \to \tilde{H}_{n-1}(X)$ is exact.

Finally since $TA/A = SA$, the exactness of

$$\begin{array}{ccccccc} \tilde{H}_n(TA) & \to & \tilde{H}_n(SA) & \xrightarrow{\partial_n} & \tilde{H}_{n-1}(A) & \to & \tilde{H}_{n-1}(TA) \\ \| & & & & & & \| \\ 0 & & & & & & 0 \end{array}$$

yields an isomorphism with $\sigma_{n-1}(A) = \partial_n^{-1}$. $TA \sim *$ so $\tilde{H}_*(TA) = 0$.

On the other hand, given $\tilde{\mathcal{H}}$ we can define $H_n(X,A) = \tilde{H}_n(X/A)$ (as before $X/\emptyset = X \cup *$ whence $H_n(X) = \tilde{H}_n(X) \oplus H_n(*) = \tilde{H}_n(X) \oplus \tilde{H}_n(S^0), S^0 = * \cup *'$). We define ∂_n by

$$\begin{array}{ccccc} H_n(X,A) & \overset{\partial_n}{\dashrightarrow} & H_{n-1}(A) = \tilde{H}_{n-1}(A) \oplus \tilde{H}_{n-1}(S^0) & & \\ \| & & & & \uparrow \\ \tilde{H}_n(X/A) & \xrightarrow{p^*} & \tilde{H}_n(SA) & \xleftarrow{\sim} & \tilde{H}_{n-1}(A) \end{array}$$

clearly

$$\begin{array}{ccccc} H_n(A) & \longrightarrow & H_n(X) & \longrightarrow & H_n(X,A) \\ \| & & \| & & \| \\ H_n(*) \oplus \tilde{H}_n(A) & \longrightarrow & H_n(*) \oplus \tilde{H}_n(X) & \longrightarrow & H_n(X/A) \end{array}$$

is exact.

We get the rest by observing that $X/A \simeq X \cup TA$ so

$$\begin{array}{ccccc} H_n(X) & \to & H_n(X \cup TA) & \to & H_n(X \cup TA/X) \\ \| & & \| & & \| \\ & & & & H_n(SA) \\ & & & & \wr\| \\ H_n(X) & \to & H_n(X/A) & \to & H_{n-1}(A) \end{array}$$

yields exact hence

$$H_n(A) \to H_n(X) \to H_n(X,A) \to H_{n-1}(A)$$

is exact.

From now on we shall deal solely with reduced generalized homology and cohomology theories. Consequently we shall neglect the word "reduced" and eliminate the ~ except for ordinary cohomology.

Let $\underline{A}$ be a spectrum. For any space X, let $k_n(X) = [\underline{S}, X \wedge \underline{A}]_n = \pi_n(X \wedge \underline{A})$ and $k^n(X) = [\underline{SX}, \underline{A}]^n = \{X, \underline{A}\}^n$.

Theorem 2.10: k^* is a cohomology theory $H^*(\ ;\underline{A})$. If $\underline{A}$ is strongly convergent, k_* is a homology theory $H_*(\ ;\underline{A})$.

Proof: It is obvious that k^* is a cohomology theory. To show that k_* is a homology theory, we investigate the "stable range" of the problem. Let $\underline{A}$ be $(r-1)$-connected. Given n, choose $N \geq n - 2r + 3$ such that for all $m \geq N$, $\varepsilon_m : SA_m \to A_{m+1}$ is $(m+n+2)$-connected and A_m is $(m+r-1)$-connected. Then for any space Y, $\pi_{n+m}(Y \wedge A_m) \simeq \pi_{n+m+1}(Y \wedge SA_m)$ since $(n+m) \leq 2(m+r) - 2$. Also $\pi_{n+m+1}(Y \wedge SA_m) \simeq \pi_{n+m+1}(Y \wedge A_{m+1})$ since ε_m is $(n+m+2)$-connected. Thus $\pi_{n+m}(Y \wedge A_m) \simeq \pi_n(Y \wedge \underline{A}) = k_n(Y)$, and $\pi_{n+m+1}(Y \wedge A_m) \cong k_{n+1}(Y)$.

Thus we have $k_n(Y) \cong \pi_{n+m}(Y \wedge A_m) \cong \pi_{n+m+1}(SY \wedge A_m) \cong k_{n+1}(SY)$. Finally if $X \subset Y$ then the cofibration $X \to Y \to Y/X$ yields the cofibration $X \wedge A_m \to Y \wedge A_m \to (Y/X) \wedge A_m$ so that

$$\pi_{m+n}(X \wedge A_m) \to \pi_{m+n}(Y \wedge A_m) \to \pi_{m+n}((Y/X) \wedge A_m)$$

is exact for $m + n < 2m + 2r - 2$ hence for $m \geq N$. This, then, yields the exactness of

$$k_n(X) \to k_n(Y) \to k_n(Y/X) .$$

Observe that for the case of $\underline{A} = \underline{K}(\pi)$ we have

$$H_n(S^0;\underline{K}(\pi)) = \begin{cases} \pi & n = 0 \\ 0 & n \neq 0 \end{cases}$$

hence $H_*(\ ;\underline{K}(\pi))$ must be regular (reduced) homology. This is a special case of the following:

Theorem 2.11: Let $T:h_* \to k_*$ be a natural transformation of homology theories (i.e. of the functors and commuting with the isomorphisms σ). Then if $T(S^0)$ is an isomorphism, so is $T(X)$ for any finite CW complex.

The proof is again the usual argument by induction on the cells using the five lemma.

There is a generalization of the ideas of Theorem 2.11 that we will find particularly useful.

Definition: A partial homology theory of bidegree (m,M) is a sequence of functors $H_n:C^* \to \mathcal{G}$ and natural transformations

$\sigma_n : H_n \to H_{n+1} \circ S$ such that

1) if $A \subset X$ satisfies HEP and A and X are $(m-1)$-connected, then $H_n(A) \to H_n(X) \to H_n(X/A)$ is exact for $n < M$ and

2) $\sigma_n(X)$ is an isomorphism if X is $(m-1)$-connected and $n < M - 1$.

Remarks:

1) This is clearly a reduced theory.

2) For each integer $m > 1$, π_* is a homology theory of bidegree $(m, 2m-1)$.

Then we extend Theorem 2.11 as follows (this proof yields a proof of Theorem 2.11, in fact):

Definition: A natural transformation $T_* : h_* \to k_*$ of partial homology theories of bidegree (m,M) is called a weak isomorphism if $T_i(X)$ is an isomorphism for $i < M$ and an epimorphism for $i = M$ for all $(m-1)$-connected finite CW complexes.

Theorem 2.12: Let $T_* : h_* \to k_*$ be a natural transformation of partial homology theories of bidegree (m,M). Assume $T_i(S^n) : h_i(S^n) \to k_i(S^n)$ is an isomorphism for all $n \geq m$ and $i < M$ and an epimorphism for $i = M$. Then T_* is a weak isomorphism.

Proof: The cofibration $X \to X \vee Y \to Y$ yields the fact that $h_i(X \vee Y) \cong h_i(X) \oplus h_i(Y)$ if X and Y are $(m-1)$-connected and $i < M$. The same is true for k_*. Thus $T_i(\bigvee S^n)$ is an isomorphism for finite wedges of n-spheres, $n \geq m$, $i < M$. Given a finite $(m-1)$-connected CW complex X, the long exact sequence for $X^m \subset X$ yields the fact that $X^m \simeq \bigvee S^m$, a finite wedge. Thus $T_i(X^m)$ is an isomorphism for $i < M$. Assume $T_i(S^r X^n)$ is an isomorphism for $i < M$ and all $r \geq 0$. Now from the cofibration $\bigvee S^n \to X^n \to X^{n+1}$ we get exact sequences connecting the maps

$$T_i(\bigvee S^{n+r}) \to T_i(S^r X^n) \to T_i(S^r X^{n+1}) \to T_{i-1}(\bigvee S^{n+r}) \to T_{i-1}(S^r X^n) .$$

Since all but the middle map are isomorphisms for $i < M$ the 5 lemma implies that $T_i(S^r X^{n+1})$ is an isomorphism for $i < M$ and all $r \geq 0$. For $i = M$, the first two are epimorphisms, and the last two are isomorphisms, $T_m(S^r X^{n+1})$ is an epimorphism. Since X is finite, $X = X^n$ for some n, hence $T_i(X)$ is an isomorphism for $i < M$ and $T_m(X)$ is an epimorphism.

CHAPTER 3. SPANIER-WHITEHEAD DUALITY

3.1 Duality Theorem

The central aim of this chapter is to show that corresponding to a finite CW complex X and a sufficiently large integer N , there is a finite CW complex $D_N(X)$, unique up to homotopy type, having many nice properties.

First $D_{N+1}(SX) \simeq D_N(X) \simeq SD_{N-1}(X)$, so that we have in a natural way a spectrum $\underline{D}X$ dual to the spectrum $\underline{S}X$.

Next we find that for each N where it is defined $H^i(D_N(X)) \simeq H_{N-i}(X)$. Finally we find that for two finite CW complexes X and Y , $[\underline{S}X,\underline{S}Y]$ may be identified, in a natural way, with $[\underline{D}Y,\underline{D}X]$.

The first step in getting this duality is to recall certain pairings of homology and cohomology.

Observe that for any abelian groups A,B , and C there is a natural "evaluation map," $e:A \otimes \mathrm{Hom}(A \otimes B,C) \to \mathrm{Hom}(B,C)$ given by $[e(a \otimes f)](b) = f(a \otimes b)$.

Given spaces X and Y and an abelian group G we can form the singular chain complexes $C_*(X)$, $C_*(Y)$ and the cochain complex $C^*(Y;G) = \mathrm{Hom}(C_*(Y),G)$. Then an evaluation map is defined

$$C_*(X) \otimes \mathrm{Hom}(C_*(X) \otimes C_*(Y),G) \to C^*(Y;G) .$$

If this is composed with the map induced by the "Eilenberg-Zilber" map (Spanier, p. 232) which gives a chain equivalence $C_*(X) \otimes C_*(Y) \rightleftarrows C_*(X \times Y)$ we have defined

$$/:C_*(X) \otimes C^*(X \times Y;G) \to C^*(Y;G)\ .$$

Specifically, if $x \in C_q(X)$, $u \in C^n(X \times Y;G)$ and $y \in C_{n-q}(Y)$, then $u/x \in C^{n-q}(Y;G)$ is given by $(u/x)(y) = u(x \times y)$. From the boundary formula on $C_*(X \times Y)$ it is immediate that $\delta(u/x) = (\delta u)/x - (-1)^{n-q}(u/\partial x)$. Thus it is easy to check that we get

$$/:H_*(X) \otimes H^*(X \times Y;G) \to H^*(Y;G)$$

induced.

Then the relative Eilenberg-Zilber (<u>Spanier</u>, p. 234) $(C_*(X) \otimes C_*(Y),\ C_*(X) \otimes C_*(B) + C_*(A) \otimes C_*(Y)) \to (C_*(X \times Y)\ ,$ $C_*(X \times B \cup A \times Y)$ for pairs (X,A), (Y,B) yields the natural transformation

$$/:H^*(X \times Y, X \times B \cup A \times Y) \otimes H_*(X,A) \to H^*(Y,B)\ .$$

If $f:(X',A') \to (X,A)$ and $g:(Y',B') \to (Y,B)$ then $g^*(u/f_*(x)) = [(f \times g)^*u]/x$ for $u \in H^*(X \times Y, X \times B \cup A \times Y)$ and $x \in H_*(X',A')$.

We shall now recall some facts about fibre bundles before we put the slant product to work.

<u>Definition</u>: (E,B,F,p) is a fibre bundle if and only if $p:E \to B$ is a map such that for every $b \in B$ there exists a neighborhood U of B and a homeomorphism $\varphi:U \times F \to p^{-1}(U)$ such that $p\varphi(u,f) = u$ for all $u \in U$. The following is proved in <u>Spanier</u>, p. 96.

<u>Lemma 3.1</u>: If (E,B,F,p) is a fibre bundle and B is paracompact and Hausdorff then p is a fibre map.

<u>Lemma 3.2</u>: Let $p:X \to B$ be a fibration. Let $X' \subset X$ be such that $p' = p|X' : X' \to B$ is a fibre map with fibre F'. Then $i_*:\pi_q(F,F') \cong \pi_q(X,X')$.

<u>Proof</u>: Let j be the composite isomorphism $\pi_q(X',F') \cong \pi_q(B) \cong \pi_q(X,F)$.

We have the triples (X,F,F') (X,X',F') leading to exact sequences (cf. <u>Spanier</u>, p. 378):

$$\begin{array}{ccccccccc} \xrightarrow{h} & \pi_r(F,F') & \xrightarrow{f} & \pi_r(X,F') & \xrightarrow{g} & \pi_r(X,F) & \xrightarrow{h} & \pi_{r-1}(F,F') \\ & & & \| & & & & \\ \xrightarrow{h'} & \pi_r(X',F') & \xrightarrow{f'} & \pi_r(X,F') & \xrightarrow{g'} & \pi_r(X,X') & \xrightarrow{h'} & \pi_{r-1}(X',F') \end{array} .$$

We wish to show that $i_* = g'f$ is an isomorphism. We know that $j = gf'$ is. Thus g is epi and f' is mono. Thus h and h' are trivial. Thus g is epi, f is mono.

Assume $i_*(x) = 0$. Then $g'(f(x)) = 0$ so $f(x) = f'(x')$ for some x'. But then $j(x') = gf'(x') = gf(x) = 0$ so $x' = 0$. Then $f(x) = f'(0) = 0$. Since f is mono $x = 0$. Thus i_* is mono.

Let $y \in \pi_r(X,X')$. Then $y = g'(z)$. j is epi so for some w $g(z) = j(w) = gf'(w)$ so $g(z - f'(w)) = 0$ so $f(x) = z - f'(w)$ for some x. But then $i_*(x) = g'f(x) = g'(z) - g'f'(w) = g'(z) = y$. Thus i_* is epi.

Let $\Delta = \{(x,x)|x \in S^{n+1}\} \subset S^{n+1} \times S^{n+1}$. Let $E = S^{n+1} \times S^{n+1} - \Delta$ and define $p:E \to S^{n+1}$ by $p(x,y) = x$. Let $F = S^{n+1} - *$.

<u>Proposition 3.3</u>: (E, S^{n+1}, F, p) is a fibre bundle.

<u>Proof</u>: Consider $S^{n+1} = \dfrac{\{z \in R^{n+1} \mid \|z\| \leq 3\} \times Z_2}{\sim}$ where $(z,0) \sim (z,1)$ if $\|z\| = 3$. We take x to be the point $(0,0)$ and we shall find a neighborhood V of x such that $p(V) \cong V \times F$. We suppress the second coordinate and consider $\{z \mid \|z\| < 3\} \subset S^{n+1}$ to contain x. Let $V = \{z \mid \|z\| < 1\}$, $D = \{z \mid \|z\| \leq 2\}$. If $(x', x'') \in V \times D$, $x' \neq x''$ then there is a unique point $z' \in R^{n+1}$ such that $\|z'\| = 2$ and x'' belongs to the closed segment from x' to z'. If $x'' = tx' + (1-t)z'$ for $t \in [0,1)$ let $h(x', x'') = (1-t)z' \neq 0$, $h(x', x') = 0$. Then define $\psi: (V \times S^{n+1}, V \times S^{n+1} - \Delta) \to V \times (S^{n+1}, S^{n+1} - x)$ by

$$\psi(x', x'') = \begin{cases} (x', x'') & x'' \notin D \\ (x', h(x', x'')) & x'' \in D \end{cases} .$$

Thus ψ gives the homeomorphism $p^{-1}(V) = V \times S^{n+1} - \Delta \cong V \times (S^{n+1} - x) \cong V \times F$.

By any rotation, what we did for x could be done for any point.

Now using Lemma 3.1 we get

<u>Theorem 3.4</u>: $p: E \to S^{n+1}$ is a fibre map.

Now we have

$$\begin{array}{ccc} S^{n+1} - * & \subset & S^{n+1} \\ \cap & & \cap \\ S^{n+1} \times S^{n+1} - \Delta & \subset & S^{n+1} \times S^{n+1} \\ \downarrow & & \downarrow \\ S^{n+1} & = & S^{n+1} \end{array}$$

where the verticals are fibre maps. Thus from Lemma 3.2

$$\pi_i(S^{n+1}, S^{n+1} - *) \cong \pi_i(S^{n+1} \times S^{n+1}, S^{n+1} \times S^{n+1} - \Delta)$$

$$\wr\!\uparrow$$

$$\pi_i(S^{n+1}) = \begin{cases} 0 & i < n+1 \\ Z & i = n+1 \end{cases}$$

Thus $H_i(S^{n+1} \times S^{n+1}, S^{n+1} \times S^{n+1} - \Delta) = \begin{cases} 0 & i < n+1 \\ Z & i = n+1 \end{cases}$

Choose $u \in H^{n+1}(S^{n+1} \times S^{n+1}, S^{n+1} \times S^{n+1} - \Delta)$ a generator.

Let $A \subset X$ be subcomplexes of S^{n+1}. Choose $X' \subset A' \subset S^{n+1}$ such that A' (resp. X') is a deformation retract of $S^{n+1} - A$ (resp. $S^{n+1} - X$).

Let $i:(X \times A', X \times X' \cup A \times A') \to (S^{n+1} \times S^{n+1}, S^{n+1} \times S^{n+1} - \Delta)$ be the inclusion. The slant product

$$/ : H^*(X \times A', X \times X' \cup A \times A') \otimes H_*(X,A) \to H^*(A',X')$$

is defined. $i^*(u) \in H^{n+1}(X \times A', X \times X' \cup A \times A')$. So there is defined

$$\gamma_u : H_q(X,A) \to H^{n+1-q}(A',X') \qquad q = 0,1,\ldots,n+1$$

by $\gamma_u(x) = i^*(u)/x$.

<u>Theorem 3.5</u>: If $A \subset X$ are subcomplexes of S^{n+1} for a fixed triangulation then γ_u is an isomorphism.

<u>Proof</u>: <u>Case 1</u>: $X = *$, $A = \emptyset$. Let $A' = S^{n+1}$, $X' = S^{n+1} - *$, then $H^{n-q+1}(S^{n+1}, S^{n+1} - *) = \begin{cases} 0 & q \neq 0 \\ Z & q = 0 \end{cases} = H_q(*)$. We need to know the

isomorphism is given by γ_u . But this follows from the fact that the map

$$H^{n+1}(S^{n+1}, S^{n+1} - *) \longleftarrow H^{n+1}(S^{n+1} \times S^{n+1}, S^{n+1} \times S^{n+1} - \Delta)$$

is an isomorphism.

Assume the theorem holds for (X,A) where $\dim(X-A) < k$, $k > 0$.

Case 2: X = k-cell, $A = \dot{X}$, the boundary of X . Then $\dot{X} = E_+ \cup E_-$ a union of two hemispheres, (k-1)-cells. $E_+ \cap E_- = \dot{E}_+ = \dot{E}_-$ a (k-2)-sphere

$$\begin{array}{ccc} H_q(X,\dot{X}) & \xrightarrow{\gamma_u} & H^{n-q+1}(\dot{X}',X') \\ \downarrow{\scriptstyle\partial} & & \downarrow{\scriptstyle\delta} \\ H_{q-1}(\dot{X},E_-) & \xrightarrow{\gamma_u'} & H^{n-q+2}(E_-',\dot{X}') \end{array}$$

commutes up to sign by a chain level formula. By induction γ_u' is an isomorphism. ∂ is an isomorphism since $H_*(X,E_-) = 0$ since both are contractible.

We need the following lemma.

Lemma 3.6: If A is a k-cell in S^n, $\tilde{H}_*(S^n - A) = 0$.

Proof: By induction on k . If $k = 0$, $A = *$ and $S^n - * \cong R^n$ is contractible. Assume the result for $k < m$, $m \geq 1$. Regard A as homeomorphic to $B \times I$, B an (m-1)-cell. $h: B \times I \to A$ the homeomorphism. Let $A' = h(B \times [0,\frac{1}{2}])$, $A'' = h(B \times [\frac{1}{2},1])$ then $A' \cup A'' = A$, $A' \cap A''$ is an (m-1)-cell. Then from the Mayer-Vietoris sequence (Spanier, p. 186) for $(S^n - A', S^n - A'')$ and from the inductive assumption that $\tilde{H}_*(S^n - A' \cap A'') = 0$ we get

$\tilde{H}_*(S^n - A) \simeq \tilde{H}_*(S^n - A') \oplus \tilde{H}_*(S^n - A'')$.

Thus if $0 \neq z \in \tilde{H}_*(S^n - A)$ then $i_*(z) \neq 0$ in $\tilde{H}_*(S^n - A_1)$ where $A_1 = A'$ or A'' . Then iterating this argument, we get a sequence of spaces

$$A \supset A_1 \supset A_2 \supset$$

and a non-zero element of $\varinjlim_i \tilde{H}_*(S^n - A_i)$. Observe that every compact set of $S^n - \cap A_i$ is contained in some $S^n - A_j$, hence (by an argument similar to that of Lemma 1.11, taking specific representative cycles for the homology classes) $H_*(S^n - \cap A_i) = \varinjlim_i \tilde{H}_*(S^n - A_i) \neq 0$. But $\cap A_i$ is an (m-1)-cell so $\tilde{H}_*(S^n - \cap A_i) = 0$ by induction. Thus we have a contradiction unless $\tilde{H}_*(S^n - A) = 0$, and the lemma is proved.

Applying this to the previous, X' and E_-' are complements of cells in S^{n+1} hence $\tilde{H}_*(X') = 0 = \tilde{H}_*(E_-')$ so $H_*(E_-', X') = 0$. Thus δ is an isomorphism, hence γ_u is also and case 2 is completed.

<u>Case 3</u>: $\dim X = k$, $A = X^{k-1}$. Let $E_1, \ldots, E_r$ be the k-simplices of $X - A$. We wish to show that $\sum_{i=1}^{r} H^*(\dot{E}_i', E_i') \simeq H^*(A', X')$ induced by the inclusion. If $r = 1$, then $(E_1, \dot{E}_1) \hookrightarrow (X, A)$ is a relative homeomorphism, hence $(S^{n+1} - A, S^{n+1} - X) \hookrightarrow (S^{n+1} - \dot{E}_1, S^{n+1} - E_1)$ is also, so the isomorphism holds for $r = 1$. The Mayer-Vietoris sequence provides the inductive step proving it for general r .

Now we have the commutative diagram

$$\begin{array}{ccc}
\sum_{i=1}^{r} H_*(E_i,\dot{E}_i) & \xrightarrow{\ \varphi\ } & H_*(X,A) \\
\Big\downarrow \gamma_u' & & \Big\downarrow \gamma_u \\
\sum_{i=1}^{r} H^*(\dot{E}_i',E_i') & \xrightarrow{\ \theta\ } & H^*(A',X')
\end{array}$$

γ_u' is an isomorphism by Case 2; φ is well known to be an isomorphism (cf. Hu [2], p. 46); θ is an isomorphism by the above; hence γ_u is an isomorphism.

Case 4: General (X,A), $\dim X = k$. Let $X_p = A \cup X^p$. By induction we will show that $\gamma_u : H_q(X_p,A) \to H^{n-q+1}(A',X_p')$ is an isomorphism. For $p = -1$, this is trivially true.

Look at the homology groups of the triad (X_p, X_{p-1}, A).

$$\begin{array}{ccccccccc}
H_{q+1}(X_p,X_{p-1}) & \to & H_q(X_{p-1},A) & \to & H_q(X_p,A) & \to & H_q(X_p,X_{p-1}) & \to & H_{q-1}(X_{p-1},A) \\
\Big\downarrow \gamma_u^1 & & \Big\downarrow \gamma_u^2 & & \Big\downarrow \gamma_u^3 & & \Big\downarrow \gamma_u^4 & & \Big\downarrow \gamma_u^5 \\
H^{n-q}(X_{p-1}',X_p') & \to & H^{n-q+1}(A';X_{p-1}') & \to & H^{n-q+1}(A';X_p') & \to & H^{n-q+1}(X_{p-1}',X_p') & \to & H^{n-q+2}(A';X_{p-1}')
\end{array}$$

Inductively γ_u^2 and γ_u^5 are isomorphisms; γ_u^1 and γ_u^4 are isomorphisms by induction for $p < k$ and by Case 3 for $p = k$. Thus γ_u^3 is an isomorphism and

$$\gamma_u : H_q(X,A) \to H^{n-q+1}(A',X')$$

is an isomorphism setting $p = k$.

In particular if $A = \emptyset$ we have

$$H_q(X) \xrightarrow[\approx]{\gamma_u} H^{n-q+1}(S^{n+1},X')$$

if $X \neq \emptyset$, $X' \subset S^{n+1} - * \subset S^{n+1}$ so the inclusion map is null-homotopic so $\tilde{H}^*(S^{n+1}) \to \tilde{H}^*(X')$ is trivial. Thus

$$H^{n-q+1}(S^{n+1},X') \approx \begin{cases} H^{n-q}(X') & q \neq 0,\ n+1 \\ Z \oplus H^n(X') & q = 0 \\ 0 & q = n+1 \end{cases}$$

Thus $\gamma_u : \tilde{H}_q(X) \approx \tilde{H}^{n-q}(X')$.

If $X \subset S^{n+1}$ and X^* is a proper deformation retract of $S^{n+1} - X$ then we pick $\alpha \in S^{n+1} - X \cup X^*$. Then we may consider $X \cup X^* \in S^{n+1} - \alpha \approx R^{n+1}$. Since $X \cap X^* = \emptyset$ we have

$$X \times X^* \subset R^{n+1} \times R^{n+1} - \Delta,\ \Delta = \{(x,x|x \in R^{n+1}\} \quad .$$

Then the deformation retraction

$$r : R^{n+1} \times R^{n+1} - \Delta \to S^n = \{(x,0) \in R^{n+1} \times R^{n+1} - \Delta \mid \|x\| = 1\}$$

given by $(x,y) \to \frac{x-y}{\|x-y\|}$ composes to give a map $f : X \times X^* \to S^n$.

If we make the assumption that X and X^* are connected based CW complexes then from the isomorphism $\tilde{H}_i(X) \approx \tilde{H}^{n-i}(X^*)$ and the other way around, we get that $H^i(X^*) = H^i(X) = 0$ for $i \geq n$. Thus $[X^*,S^n] = [X,S^n] = 0$ from Theorem 2.6 so the composite $X \vee X^* \to X \times X^* \to S^n$ is null-homotopic thus a map (unique if $[SX^*,S^n] = 0 = [SX,S^n]$)

$$u : X \wedge X^* \to S^n$$

is defined.

<u>Claim</u>: For $\iota_n \in H^n(S^n)$, $u^*(\iota_n)/:\tilde{H}_q(X) \simeq \tilde{H}^{n-q}(X^*)$.

This follows from the commutativity of the following horrible diagram. (Represent R^{n+1} as $S^{n+1} - \alpha$ as before.)

$$H^{n+1}(S^{n+1} \times S^{n+1}, S^{n+1} \times S^{n+1} - \Delta) \to H^{n+1}(X \times S^{n+1}, X \times (S^{n+1} - X))$$

$$\downarrow \approx$$

$$H^{n+1}(R^{n+1} \times S^{n+1}, R^{n+1} \times S^{n+1} - \Delta) \nearrow \qquad \approx$$

$$\downarrow \approx$$

$$H^{n+1}(R^{n+1} \times R^{n+1}, R^{n+1} \times R^{n+1} - \Delta) \to H^{n+1}(X \times S^{n+1}, X \times X^*)$$

$$\approx \Big\uparrow \delta \qquad\qquad \Big\uparrow \delta$$

$$H^n(R^{n+1} \times R^{n+1} - \Delta) \xrightarrow{\hspace{3cm}} H^n(X \times X^*)$$

$$\approx \Big\downarrow \qquad \nearrow$$

$$H^n(S^n) \xrightarrow{\hspace{2cm}f^*}$$

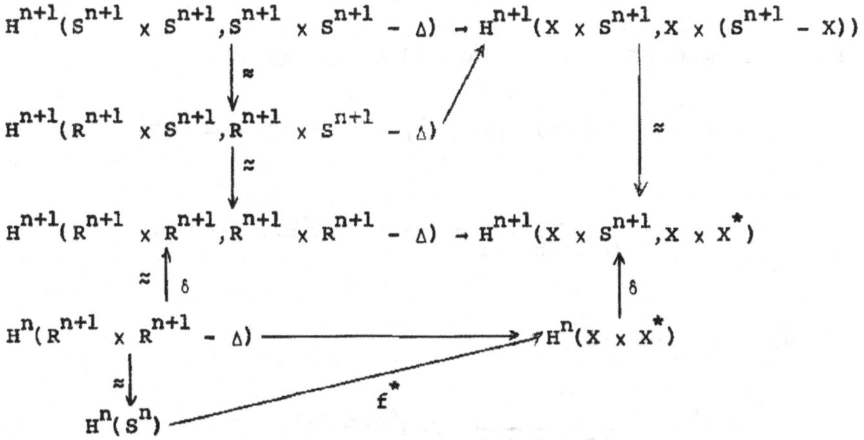

All unmarked maps are inclusions.

<u>Definition</u>: If $i:A \subset B$ then A is an S-(deformation) retract of B if and only if there exists $j \in \{B,A\}$ such that $j \circ \{i\} = \{1\}$ (and $\{i\} \circ j = \{1\}$) .

Observe that if $i:A \hookrightarrow B$ is an S-deformation retract, then $\underline{i}:\underline{S}A \to \underline{S}B$ is a homotopy equivalence. Thus $\underline{i}_*:H_*(\underline{S}A) \to H_*(\underline{S}B)$ is an isomorphism. But this is the same as $i_*:\tilde{H}_*(A) \to \tilde{H}_*(B)$. Thus i induces a homology isomorphism. Thus, in particular, if A and B are 1-connected CW complexes, i is a homotopy equivalence.

<u>Definition</u>: If X, X^* are based CW complexes, then X^* is a

geometric n-dual of X if X and X^* can be embedded in S^{n+1} so that X^* is an S-deformation retract of $S^{n+1} - X$.

X^* is an n-dual of X if and only if there is a map $u: X \wedge X^* \to S^n$ such that $u^*(\iota_n)/ : \tilde{H}_q(X) \to \tilde{H}^{n-q}(X^*)$ is an isomorphism, where ι_n generates $H^n(S^n)$.

u is called an n-duality map.

<u>Lemma 3.7</u>: If $u: X \wedge X^* \to S^n$ is an n-duality map then

$$(SX) \wedge X^* \longrightarrow S(X \wedge X^*) \xrightarrow{Su} S(S^n) \longrightarrow S^{n+1}$$

and

$$X \wedge (SX^*) \xrightarrow[\text{interchange}]{} S(X \wedge X^*) \xrightarrow{Su} S(S^n) \to S^{n+1}$$

are (n+1)-duality maps and

$$X^* \wedge X \xrightarrow[\text{interchange}]{} X \wedge X^* \xrightarrow{u} S^n$$

is an n-duality map.

<u>Proof</u>: The first comes from the fact that

$$u^*(\iota_n)/ : \tilde{H}_q(X) \to \tilde{H}^{n-q}(X^*)$$

so

$$u^*(\iota_n)/ \in \mathrm{Hom}(\tilde{H}_q(X), \tilde{H}^{n-q}(X^*))$$

and the map $H^n(X \wedge X^*) \to \tilde{H}^q(X; \tilde{H}^{n-q}(X^*)$

$$\downarrow$$

$$\mathrm{Hom}(\tilde{H}_q(X), \tilde{H}^{n-q}(X^*))$$

takes $u^*(\iota_n) \mapsto u^*(\iota_n)/$.

Then it is clear that the new maps are in fact isomorphisms. The final statement, follows from the universal coefficient theorem.

3.2 Duality in Certain Spectra and Maps

Let us assume that we have a duality map $u: X^* \wedge X \to S^n$. Then we also have duality maps $S^k u: (S^k X^*) \wedge X \to S^{n+k}$. This leads to maps $f_k = \widetilde{S^k u}: S^k X^* \to F(X, S^{n+k})$ the adjoints. By commutativity of

$$\begin{array}{ccc} S(S^k X^*) & \xrightarrow{Sf_k} & SF(X, S^{n+k}) \\ \downarrow & & \downarrow \\ S^{k+1} X^* & \xrightarrow{f_{k+1}} & F(X, S^{n+k+1}) \end{array}$$

we have induced $f: \underline{S}X^* \to \underline{F}(X, \underline{S})$. Reindex so that $S^k X^*$ is the $(n+k)$-th term of $\underline{S}X^*$.

Theorem 3.8: f is a w.h.e.

Proof: It suffices to prove that f induces a homology isomorphism. The idea behind the proof is represented by the following: $H_*(\underline{F}(X,\underline{S}) \cong \pi_*(\underline{F}(X,\underline{S}) \wedge \underline{K}) \cong \pi_*(\underline{F}(X,\underline{K})) \cong [\underline{S}X, \underline{K}] \cong H^*(X) \cong H_*(X^*) \cong H_*(\underline{S}X^*)$, where $\underline{K} = \underline{K}(Z)$. We shall prove these isomorphisms in several stages.

Look at the map $\rho: F(X,Y) \wedge Z \to F(X, Y \wedge Z)$ given by $\rho(f \wedge z)(x) \cong f(x) \wedge z$.

Theorem 3.9: Let X, Y and Z be CW spaces with $\dim X \leq k$, Y $(n-1)$-connected and Z $(m-1)$-connected. Then $\rho: F(X,Y) \wedge Z \to F(X, Y \wedge Z)$ is $(2n-2k+m)$-connected.

Proof: Fix X and Y. Then let $h_i(Z) = \pi_i(F(X,Y) \wedge Z)$ and $k_i(Z) = \pi_i(F(X, Y \wedge Z))$. We claim that h_* and k_* are partial homology theories of bidegree $(m, 2(m+n-k-1))$ and that $\rho: h_* \to k_*$ is an isomorphism in the appropriate degrees.

First observe that $\pi_i(F(X,Y)) \simeq [S^iX,Y] = 0$ for $i + k \leq n - 1$ so $F(X,Y)$ is $(n-k-1)$-connected. Then if $A \subset Z$ are $(m-1)$-connected spaces with the HEP, $F(X,Y) \wedge A \to F(X,Y) \wedge Z \to F(X,Y) \wedge Z/A$ is a cofibration of $(n+m-k-1)$-connected spaces. Hence there is an exact homotopy sequence (Theorem 1.6) starting with $\pi_{2(n+m-k-1)}(F(X,Y) \wedge A) = h_{2(n+m-k-1)}(A)$. The isomorphism $h_i(Z) \simeq h_{i+1}(SZ)$ is evident from the exact sequence.

Next observe that $k_i(Z) \simeq [S^iX, Y \wedge Z]$. Now $Y \wedge A \to Y \wedge Z \to Y \wedge Z/A$ is a cofibration of $(n+m-1)$-connected spaces, hence from Theorem 1.16, there is an exact sequence beginning $[S^iX, Y \wedge A] \cong k_i(A)$ for $i + k \leq 2(n+m-1)$, hence, for $i \leq 2(n+m-k-1)$. So h_* and k_* are partial homology theories of bidegree $(m, 2n-2k+m-1)$. (Actually they can both be extended a bit.)

Now let $Z = S^t$ $t \geq m$. The following diagram clearly commutes up to sign.

$$\begin{array}{ccccc}
h_i(S^t) & & \longrightarrow & & k_i(S^t) \\
\| & & & & \| \\
\pi_i(F(X,Y) \wedge S^t) & & \xrightarrow{\rho_*} & & \pi_i(F(X,Y \wedge S^t)) \\
\uparrow & & & & \uparrow\wr \\
\pi_{i-t}(F(X,Y)) & \cong & [S^{i-t}X,Y] & \longrightarrow & [S^iX,S^tY]
\end{array}$$

Left vertical map: iso for $(i-t) < 2(n-k)-1$, epi for $(i-t) \leq 2(n-k)-1$ (Theorem 1.9).

Bottom map: isomorphism for $(i-t+k) < 2n-1$, epimorphism for $(i-t+k) \leq 2n-1$ (Theorem 1.21).

Thus ρ_* is an isomorphism for $i < 2n - 2k + t$ and an epimorphism for $i = 2n - 2k + t$. Thus for $t \geq m$, ρ_* is an isomorphism for $i < 2n - 2k + m - 1$ and an epimorphism for

$i = 2n - 2k + m - 1$. Thus by Theorem 2.12, for any $(m-1)$-connected Z , $\rho_*: h_i(Z) \to k_i(Z)$ is also, hence $\rho: F(X,Y) \wedge Z \to F(X, Y \wedge Z)$ is $(2n-2k+m)$-connected.

There is a very useful principle which will be demonstrated in Corollary 3.11. It is this: If $f: Y \to Z$ is n-connected and X is a CW space of dimension m , then $F(X,f): F(X,Y) \to F(X,Z)$ is $(n-m)$-connected. This is immediate from the observation that the map $\pi_i(F(X,f)): \pi_i(F(X,Y)) \to \pi_i(F(X,Z))$ is equivalent to $[S^i X, f]: [S^i X, Y] \to [S^i X, Z]$ which is an isomorphism for $i + m < n$ and an epimorphism for $i + m = n$.

<u>Corollary 3.10</u>: Let X be $(n-1)$-connected. Then $\varphi: X \to \Omega^r S^r X$, the adjoint of the identity map of $S^r X$, is $(2n-1)$-connected, and $\theta: S^r \Omega^r X \to X$, the adjoint of the identity map of $\Omega^r X$, is $(2n-r)$-connected.

<u>Proof</u>: We have already proved that φ is $(2n-1)$-connected in Corollary 1.23. The following diagram commutes

$$\begin{array}{ccc} S^r \Omega^r X & \xrightarrow{\theta} & X \\ \| & & \\ S^r F(S^r, X) & & \downarrow \varphi \\ \rho \downarrow & & \\ F(S^r, S^r X) & = & \Omega^r S^r X \end{array}$$

ρ is $(2n-r)$-connected by Theorem 3.9 and φ is $(2n-1)$-connected. Thus θ is $(2n-r)$-connected.

<u>Corollary 3.11</u>: If X is a CW space of dimension $\leq k$, then the composite $F(X,S^n) \wedge K(Z,m) \xrightarrow{\rho} F(X,S^n K(Z,m)) \cong F(X,S^n \Omega^n K(Z,n+m)) \xrightarrow{F(X,\theta)}$ $F(X,K(Z,n+m))$ is $\min\{2n+m-2k,\ 2m+n-k\}$-connected. In particular $F(X,S^m) \wedge K(Z,m) \to F(X,K(Z,2m))$ is $(3m-2k)$-connected.

<u>Proof</u>: ρ is $(2n-2k+m)$-connected. θ is $[2(n+m)-n]$-connected, so $F(X,\theta)$ is $(2m+n-k)$-connected; and the result is immediate.

But $\pi_i(F(X,K(Z,2m)) \cong [S^i X, K(Z,2m)] \cong H^{2m}(S^i X) \cong H^{2m-i}(X)$.

Thus

<u>Corollary 3.12</u>: $\pi_i(F(X,S^m) \wedge K(Z,m)) \cong H^{2m-i}(X)$ for $i < 3m - 2k$ where $\dim X \leq k$.

<u>Theorem 3.13</u>: The composite

$$\pi_{i+m}(X \wedge K(Z,m)) \xrightarrow{h} H_{i+m}(X \wedge K(Z,m)) \cong \Sigma\, \tilde{H}_j(X; \tilde{H}_{i+m-j}(K(Z,m)) \longrightarrow \tilde{H}_i(X)$$

$$\varphi_i(X): \pi_{i+m}(X \wedge K(Z,m)) \to \tilde{H}_i(X)$$

is an isomorphism for $0 < i \leq m + t - 1$, where X is a $(t-1)$-connected CW complex.

<u>Proof</u>: Let $h_i(X) = \pi_{i+m}(X \wedge K(Z,m))$. Then for any t, h_* is a partial homology theory of bidegree $(t,m+t)$. Then $\varphi_i : h_i \to H_i(\ \ )$ is a natural transformation of partial homology theories of bidegree $(t,m+t)$. It suffices to show that $\varphi_i(S^k)$ is an isomorphism for $k \geq t$ and $i \leq m + k$. But by Corollary 3.10 $S^k K(Z,m) \to K(Z,m+k)$ is $(2m+k)$-connected. Thus for $0 < i < m+k$, $\pi_{i+m}(S^k K(Z,m)) = 0$

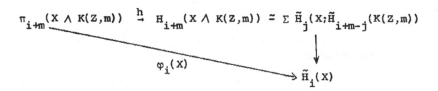

except for $i = k$ and $\pi_{m+k}(S^k K(Z,m)) = Z$. But the same is true of $H_i(S^k)$ and the Hurewicz isomorphism theorem yields the isomorphism for $i = k$.

We proved in Corollary 3.12 that $\pi_{i+m}(F(X,S^m) \wedge K(Z,m)) \approx \tilde{H}^{m-i}(X)$ for $i < 2m - 2k$, $\dim X \leq k$. Now we have that $\pi_{i+m}(F(X,S^m) \wedge K(Z,m)) \cong \tilde{H}_i(F(X,S^m))$ for $i \leq m + m - k - 1$. Thus

<u>Theorem 3.14</u>: If $\dim X \leq k$ then for $i \leq 2m - 2k - 1$

$$\tilde{H}_i(F(X,S^m)) \cong \tilde{H}^{m-i}(X) .$$

Recall the map $u: X^* \wedge X \to S^m$ which we have assumed induces $u^*(\iota_n)/:\tilde{H}_i(X^*) \cong H^{m-i}(X)$. It defines $\tilde{u}: X^* \to F(X,S^m)$ by $\tilde{u}(x^*)(x) = u(x^* \wedge x)$.

<u>Lemma 3.15</u>: The diagram

$$\begin{array}{ccc} \tilde{H}^{m-i}(X) & \cong & \tilde{H}_i(F(X,S^m)) \\ \wr\uparrow & \nearrow_{\tilde{u}_*} & \\ \tilde{H}_i(X^*) & & \end{array}$$

commutes.

<u>Proof</u>: Recall the evaluation map $e: F(X,S^m) \wedge X \to S^m$. The diagram

$$\begin{array}{ccc} X^* \wedge X & \xrightarrow{u} & S^m \\ {\scriptstyle \tilde{u}\wedge 1}\downarrow & \nearrow_{e} & \\ F(X,S^m) \wedge X & & \end{array}$$

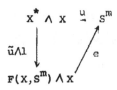

obviously commutes: $e(\hat{u} \wedge 1)(x^* \wedge x) = e(\tilde{u}(x^*) \wedge x) = \tilde{u}(x^*)(x) = u(x^* \wedge x)$. Thus by naturality of the Künneth formula

$$\begin{array}{ccccc} \tilde{H}_i(X^*) \otimes \tilde{H}_{m-i}(X) & \overset{\alpha}{\to} & \tilde{H}_m(X^* \wedge X) & \xrightarrow{u_*} & H_m(S^m) = Z \\ \Big\downarrow {\scriptstyle \tilde{u}\otimes 1} & & & \nearrow {\scriptstyle e_*} & \\ \tilde{H}_i(F(X,S^m)) \otimes \tilde{H}_{m-i}(X) & \overset{\beta}{\to} & \tilde{H}_m(F(X,S^m) \wedge X) & & \end{array}$$

commutes.

The map $u_*\alpha$ is the same as the composite

$$\begin{array}{ccl} \tilde{H}_i(X^*) \otimes \tilde{H}_{m-i}(X) & \cong & \tilde{H}^{m-i}(X) \otimes \tilde{H}_{m-i}(X) \\ & & \Big\downarrow {\scriptstyle p \otimes 1} \\ & & \mathrm{Hom}(\tilde{H}_{m-i}(X), Z) \otimes \tilde{H}_{m-i}(X) \to Z \ . \end{array}$$

Thus in order to show that $\tilde{u}$ is an isomorphism for i sufficiently small compared to m , it will suffice to show that the following diagram commutes up to sign:

$$\begin{array}{ccc} \tilde{H}^{m-i}(X) \otimes \tilde{H}_{m-i}(X) & \xrightarrow{\sim} & \tilde{H}_i(F(X,S^m)) \otimes \tilde{H}_{m-i}(X) \\ \Big\downarrow {\scriptstyle p\otimes 1} & \searrow {\scriptstyle \varphi(X)} \quad \searrow {\scriptstyle \theta(X)} & \Big\downarrow {\scriptstyle e_*\beta} \\ \mathrm{Hom}(\tilde{H}_{m-i}(X), Z) \otimes \tilde{H}_{m-i}(X) & \longrightarrow & Z \end{array}$$

i.e. that $\theta(X) = \pm \varphi(X)$.

First define the adjoint maps $\tilde{\theta}(X)$, $\tilde{\varphi}(X) : \tilde{H}^{m-i}(X) \to \mathrm{Hom}(H_{m-i}(X), Z)$. By the adjointness it follows that $\tilde{\varphi}(X)$ is simply the projection $p:\tilde{H}^{m-i}(X) \to \mathrm{Hom}(\tilde{H}_{m-i}(X),Z)$. Thus we have $\tilde{\theta}(X):\tilde{H}^{m-i}(X) \to \mathrm{Hom}(\tilde{H}_{m-i}(X),Z)$ defined and natural in X and we wish to show that it is $\pm p$.

We can use the fact that $\tilde{H}^{m-i} = [\ ,K(Z,m-i)]$. But then $\tilde{\theta}$ is determined by some element of $\mathrm{Hom}(\tilde{H}_{m-i}(K(Z,m-i)),Z) \cong Z$. $\tilde{\varphi}$ is obviously determined by a generator. Thus $\tilde{\theta} = \lambda\tilde{\varphi}$ for some integer λ . But $\tilde{\theta}(S^o)$ is clearly an isomorphism so $\lambda = \pm 1$ and the lemma is proved.

Thus $\tilde{u}_*:H_i(X^*) \to H_i(F(X,S^m))$ is an isomorphism for $i \leq 2m - 2k - 1$ where $\dim X \leq k$. But then if we look at $f_q:S^qX^* \to F(X,S^{m+q})$, $\tilde{H}_i(f_q)$ is an isomorphism for $i \leq 2m - 2k - 1 + 2q$. Thus if we pass to the map of spectra $f:\underline{S}X^* \to F(X,\underline{S})$ (reindexed so that S^qX^* is the (m+q)-th space) we have that f induces a homology isomorphism. $\underline{S}X^*$ is strongly convergent. Since $H_i(f_q)$ is an isomorphism for $i \leq 2m - 2k - 1 + 2q$, $F(X,S^{m+q})$ is (q-1)-connected and Lemma 3.9 says that $SF(X,S^{m+q}) \to F(X,S^{m+q+1})$ is $2(m+q) - 2k$ connected so $F(X,\underline{S})$ is strongly convergent. Then applying Theorem 2.7 we get

Theorem 3.8: $f:\underline{S}X^* \to F(X,\underline{S})$ is a weak homotopy equivalence.

Observe that for A and B finite CW complexes $\{A \wedge B,C\} \cong [S^NA \wedge B,S^NC] \cong [S^NA,F(B,S^NC)]$ for N sufficiently large. Thus letting $N \to \infty$ we get $\{A \wedge B,C\} \cong \{A,\underline{F}(B,\underline{S}C)\}$. Let X and Y be finite CW complexes embedded in S^{m+1} and X^* and Y^* be their m-duals. Then using this, the above theorem, and the fact that $Y^{**} \cong Y$ we get that

$$\{X,Y\} \cong \{X,\underline{S}Y\} \cong \{X,F(Y^*,\underline{S})\} \cong \{X \wedge Y^*,S^0\} \cong \{Y^*,F(X,\underline{S})\} \cong \{Y^*,X^*\} .$$

Thus, given duality maps $u:X^* \wedge X \to S^m$ and $v:Y^* \wedge Y \to S^m$ we get an isomorphism $D(u,v):\{X,Y\} \cong \{Y^*,X^*\}$ where $f:S^r X \to S^r Y$ corresponds to $g:S^t Y^* \to S^t X^*$ if and only if the following diagram is stably commutative (i.e. some suspension of it commutes):

$$
\begin{array}{ccc}
S^t Y^* \wedge S^r X & \xrightarrow{\;g \wedge 1\;} & S^t X^* \wedge S^r X \\
{\scriptstyle 1 \wedge f}\Big\downarrow & & \Big\downarrow{\scriptstyle {}^t u^r} \\
S^t Y^* \wedge S^r Y & \xrightarrow[\;{}^t v^r\;]{} & S^{m+t+r}
\end{array}
$$

where ${}^t u^r$ and ${}^t v^r$ are the appropriate suspensions of u and v .

Thus duality is almost an isomorphism of the stable category with itself. Unfortunately it isn't quite because of the choices involved. For most practical purposes, however, we can regard it as a functor assigning to spaces X embeddable in S^n , the space $D_n X$ in such a way that $D_n D_n X = X$ and $\{X,Y\} \cong \{D_n Y,D_n X\}$.

Recall that if $X \wedge D_n X \to S^n$ is a duality map then so are $(S^m X) \wedge D_n X \to S^{n+m}$ and $X \wedge S^m D_n X \to S^{n+m}$. Thus $D_{n+m}(S^m X) = D_n X$ and $D_{n+m} X = S^m D_n X$.

We can make the category C of finite CW complexes and S-maps a graded category by defining

$$\{X,Y\}_n = \{S^n X,Y\} = [\underline{S}X,\underline{S}Y]_n .$$

Then

$$\{X,Y\}_n = \{S^n X,Y\} \cong \{D_{m+n}Y,D_{m+n}S^n X\} = \{S^n D_m Y,D_m X\} = \{D_m Y,D_m X\}_n .$$

Thus D_m preserves grading.

Observe that if $X \to Y \to C_f$ is cofibration then we get long exact sequences

$$\{W,X\}_n \to \{W,Y\}_n \to \{W,C_f\}_n \to \{W,X\}_{n-1}$$

$$\{X,W\}_n \leftarrow \{Y,W\}_n \leftarrow \{C_f,W\}_n \leftarrow \{X,W\}_{n+1}$$

Also if $X \to Y \to C_f$ is cofibration, then

$$D_n C_f \to D_n Y \to D_n X$$

acts like one in terms of the long exact sequences.

This leads to being able to consider certain problems by only looking at dual problems.

Example: Freyd conjectures (Freyd [2]) the following: If X and Y are finite CW complexes and $f \in \{X,Y\}$ is such that $f_*:\pi_*^S(X) \to \pi_*^S(Y)$ is zero then $f = 0$. $(\pi_*^S(\ ) = H_*(\ ;\underline{S}))$.

A dual conjecture replaces f_* by $f^*:\{Y,S^0\}_* \to \{X,S^0\}_*$.

They are equivalent: Assume the former conjecture true. Pick $f \in \{X,Y\}$ with $f^*:\{Y,S^0\} \to \{X,S^0\}$ zero; then for N large we have $D_N f \in \{D_N Y, D_N X\}$ with $(D_N f)_*:\{S^N,Y\}_* \to \{S^N,X\}_*$ zero hence $D_N f = 0$ hence $f = 0$. Similarly the second conjecture implies the first.

We now consider the following category $\mathcal{S}$: the objects are pairs (X,n) where X is a finite CW space and a given embedding of it into some sphere (and hence all higher spheres) and n is an integer and we set $(X,n) = (SX,n-1)$. The morphisms are $\mathcal{S}((X,n),(Y,m)) = \{S^{r+n}X, S^{r+m}Y\}$ where $r+n,\ r+m \geq 0$. Observe that we can use the fact that

1. This is independent of r

2. This is unaffected by replacement of (X,n) by $(SX,n-1)$.

Roughly, the objects of $\mathcal{S}$ are finite CW complexes and their formal desuspensions. We write $(X,n) = S^n X$ for any integer n and this makes sense in $\mathcal{S}$. We can always desuspend objects in $\mathcal{S}$. Call an object of $\mathcal{S}$ "real" if it is equal to some $(X,0)$.

$\mathcal{S}$ has an advantage when it comes to Spanier-Whitehead duality: we can talk about <u>the</u> dual as follows:

Given a real object X choose n sufficiently large that $D_n X$ exists. Then as usual $D_{n+1} X = SD_n X$ so if we define $DX = S^{-n} D_n X$ we get a unique object of $\mathcal{S}$ (in general, it is not real), independent of n . For any object of $\mathcal{S}$ we extend by setting $DS^r X = S^{-r} DX$. This also is unique up to homotopy type since

$$SD_n SX \simeq D_n X$$

whenever $D_n X$ is defined. Observe also that the maps $e^n_X : X \wedge D_n X \to S^n$ will yield in $\mathcal{S}$ a natural map $e_X : X \wedge DX \to S^0$ by taking desuspension.

<u>Theorem 3.17</u>: D acts like a contravariant functor which is an anti-automorphism and involution on $\mathcal{S}$. I.e.,

1) $D^2 X \simeq X$

2) $\mathcal{S}(X,Y) \simeq \mathcal{S}(DY,DX)$

We observe that $DS^n \simeq S^{-n}$ for all n . Also observe that since $D_n X \wedge D_m Y = D_{n+m}(X \wedge Y)$ for spaces we get $DX \wedge DY = D(X \wedge Y)$ in $\mathcal{S}$.

<u>Theorem 3.18</u>: $\mathcal{S}(X \wedge Y, Z)$ is naturally equivalent to $\mathcal{S}(X, DY \wedge X)$. Thus $\wedge Y$ and $DY\wedge$ are adjoint functors.

<u>Proof</u>: First observe that it will suffice to prove this for real objects Y since if $Y = S^r Y'$ then $\mathcal{S}(X \wedge Y, Z) = \mathcal{S}(S^r X \wedge Y', Z) = \mathcal{S}(X \wedge Y', S^{-r} Z)$ and $\mathcal{S}(X, DY \wedge Z) = \mathcal{S}(X, DS^r Y' \wedge Z) = \mathcal{S}(X, (S^{-r} DY') \wedge Z) = \mathcal{S}(X, DY' \wedge S^{-r} Z)$.

Fixing X and Z define $k^n(Y) = \mathcal{S}_{-n}(X, DY \wedge Z)$ and $\bar{k}^n(Y) = \mathcal{S}_{-n}(X \wedge Y, Z)$. Observe that k^* and $\bar{k}^*$ are cohomology theories. Define $\alpha : k^* \to \bar{k}^*$ as the composite

$$\begin{array}{ccc}
\mathcal{S}_{-n}(X, DY \wedge Z) & \xrightarrow{Y\wedge -} & \mathcal{S}_{-n}(Y \wedge X, Y \wedge DY \wedge Z) \\
\Big\downarrow \alpha(Y) & & \Big\downarrow \mathcal{S}_{-n}(Y \wedge X, e_Y \wedge Z) \\
\mathcal{S}_{-n}(X \wedge Y, Z) & \xleftarrow{T^*} & \mathcal{S}_{-n}(Y \wedge X, S^0 \wedge Z)
\end{array}$$

where $T : X \wedge Y \to Y \wedge X$ is the twist and $S^0 \wedge Z$ is identified with Z . It suffices to show that $\alpha(S^0)$ is an isomorphism, but this is trivial since each of the three maps is then an isomorphism.

Observe that reduced homology and cohomology can be defined on by setting $\tilde{H}_k(S^r X) = \tilde{H}_{k-r}(X)$, $\tilde{H}^k(S^r X) = \tilde{H}^{k-r}(X)$ and we have for $\tilde{H} = \tilde{H}(\ ; Z)$

$$\tilde{H}_{-k}(DX) = \tilde{H}_{-k}(S^{-n} D_n X) = \tilde{H}_{n-k}(D_n X) \cong \tilde{H}^k(X)$$

for X real and hence by suspension for all X .

Observe that Theorem 3.18 makes sense and holds for any space Z .

Setting $X = S^n$ yields $\mathcal{S}_n(Y,Z) = \mathcal{S}(S^nY,Z) \simeq \mathcal{S}(S^n, DY \wedge Z) =$ $\pi_n^s(DY \wedge Z)$.

Next observe that if $\underline{A}$ is a convergent spectrum then $H_k(W;\underline{A}) =$ $\varinjlim_r \pi_{k+r}(W \wedge A_r) = \varinjlim_r \pi_{k+r}^s(W \wedge A_r)$. Thus $H_k(DY;\underline{A}) =$ $\varinjlim_r \pi_{k+r}^s(DY \wedge A_r) = \varinjlim_r \mathcal{S}(S^{k+r}Y, A_r) = \varinjlim_r [S^{k+r}Y, A_r] =$ $\varinjlim_r [S^rY, A_{r-k}] = H^{-k}(Y;\underline{A})$.

Thus duality holds for homology with coefficients in a convergent spectrum. Also if k^* is any (reduced) cohomology theory, we get a dual homology theory by defining $k_n(X) = k^{-n}(DX)$.

Finally let us observe that if we take the category of spectra, take the full subcategory $\mathcal{S}'$ of all $(\underline{S}X)^d$ where X is a finite CW space then $\mathcal{S}' \simeq \mathcal{S}$ ($S^nX \in \mathcal{S}$ corresponds to $(\underline{S}X)^n$).

Incidentally we may ask about homology and cohomology theories defined in $\mathcal{S}$ in general and ask if they come from spectra.

Theorem 3.19: If $\mathcal{H}$ is a generalized homology theory defined on $\mathcal{S}$ and $H_n(S^0)$ is countable for all n , then there exists a spectrum $\underline{A}$ such that $\mathcal{H} = H_*(\ ;\underline{A})$.

This theorem is due to E. H. Brown, Jr. (cf. Brown) and we shall not prove it.

CHAPTER 4. STABLE HOMOTOPY IN CATEGORICAL ALGEBRA

4.1 The Freyd Category

We wish to consider some of the more algebraic properties of the category $\mathcal{S}$; properties which we study from the point of view of Freyd [2] and [3]. Observe that $\mathcal{S}$ is additive; i.e. $\mathcal{S}(X,Y)$ is a group. We have constructions $\wedge$ and $\vee$ and mapping cones.

$\vee$ defines finite product and coproduct: i.e.

$$\mathcal{S}(X \vee X',Y) = \mathcal{S}(X,Y) \oplus \mathcal{S}(X',Y),\ \mathcal{S}(X,Y \vee Y') = \mathcal{S}(X,Y) \oplus (X,Y')\ ;$$

the first is true even unstably, the second follows from Theorem 1.18.

We observe that $\mathcal{S}$ has the property that given any map $f:X \to Y$, we have $\sigma_f:S^{-1}C_f \to X$ defined in $\mathcal{S}$ making $S^{-1}C_f \to X \to Y$ a cofibration. Furthermore we observe that $0:A \to B$ has mapping cone $B \vee SA$. Thus if a map $f:X \to Y$ splits: i.e. if there is a map $g:Y \to X$ with $gf = 1_X$, then we form the cofibration $S^{-1}C_f \xrightarrow{\sigma} X \xrightarrow{f} Y$. Then $\sigma = 1_X\sigma = gf\sigma = g(f\sigma) = 0$. Thus $Y = X \vee S(S^{-1}C_f) = X \vee C_f$. Thus in $\mathcal{S}$, splitting also means the smaller space is a (wedge) factor of the bigger.

Observe that if $X = S^nX'$ with X' real we can define $\dim X = n + \dim X'$ and we can say X is $(r+n)$-connected if X' is r-connected. We can define $\pi_r(X) = \mathcal{S}(S^r,X)$ for every integer r and define $(r+n)$-connected as usual. These definitions agree. Then $\mathcal{S}(X,Y) = 0$ if $\dim X \leq r$ and Y is r-connected. (Extending Lemma 1.14)

Thus for X and Y fixed $\mathcal{S}(S^nX,Y) = 0$ if $n \leq r - \dim X$ where Y is r-connected.

For each object X we have graded rings $\mathrm{End}_*(X) = \mathcal{S}_*(X,X)$. The unit $1 \in \mathcal{S}_o(X,X)$ is the identity. In general these rings are not commutative or graded commutative (that is $ab = -ba$ if a and b are of odd degree). $\mathrm{End}_*(S^n) = \pi_*(\underline{S})$ is the stable homotopy ring of the sphere. It <u>is</u> graded commutative and is in fact the <u>only</u> such case. $\mathrm{End}\ X$ is the subring $\mathrm{End}_o(X)$.

Now, additive categories are nice, but there is a much nicer restriction of them called abelian categories: each map has a kernel, cokernel and image. Exactness makes sense. One can do a great deal with abelian categories. We are going to put $\mathcal{S}$ inside a certain abelian category. Note: A diagram in an abelian category will be called <u>exact</u> if every row and column is exact and the diagram commutes.

Recall the following definitions in an abelian category A :

<u>Definition</u>: P is a projective object of A if and only if given the exact diagram

$$\begin{array}{ccccc} & & P & & \\ & & \downarrow & & \\ M & \rightarrow & N & \rightarrow & 0 \end{array}$$

there is a map $P \to M$ making the diagram commute.

Q is an injective object if and only if given the exact diagram

$$\begin{array}{ccccc} & & Q & & \\ & & \uparrow & & \\ M & \leftarrow & N & \leftarrow & 0 \end{array}$$

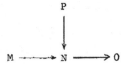

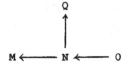

there is a map $M \to Q$ making the diagram commute. Equivalent definitions are the requirements: epimorphisms onto P and monomorphisms from Q split.

We say A has enough projectives (resp. injectives) if and only if for every object M there is a projective P (an injective Q) and an exact sequence $P \to M \to 0$ $(0 \to M \to Q)$.

<u>Definition</u>: A frobenius category is an abelian category in which projectives and injectives are the same and in which there are enough projectives and injectives.

We are going to define a frobenius category $\mathfrak{F}$ called the Freyd category in which $\mathcal{S}$ will generate the full subcategory of projective-injectives. (I.e., every projective-injective is isomorphic to something in $\mathcal{S}$.)

An object α of $\mathfrak{F}$ is to be a morphism of $\mathcal{S}$, say $\alpha \in \mathcal{S}(X,Y)$. If $\alpha \in \mathcal{S}(X,Y)$ and $\alpha' \in \mathcal{S}(X',Y')$ then a morphism $m \in \mathfrak{F}(\alpha,\alpha')$ is a pair (m',m'') $m' \in \mathcal{S}(X,X')$, $m'' \in \mathcal{S}(Y',Y'')$ with $m''\alpha = \alpha'm'$, i.e.

$$\begin{array}{ccc} X & \xrightarrow{m'} & X' \\ \alpha\big\downarrow & & \big\downarrow\alpha' \\ Y & \xrightarrow{m''} & Y' \end{array}$$

commutes, subject to the identification

$$\begin{array}{ccc} (A' & \to & A) \\ f'\big\downarrow & & \big\downarrow f \\ (B' & \to & B) \end{array} \quad = \quad \begin{array}{ccc} (A' & \to & A) \\ g'\big\downarrow & & \big\downarrow g \\ (B' & \to & B) \end{array}$$

if and only if $A' \to A \xrightarrow{f} B = A' \to A \xrightarrow{g} B$ (hence if and only if $A' \xrightarrow{f'} B' \to B = A' \xrightarrow{g'} B' \to B)$.

We define a functor $\varphi:\mathfrak{S} \to \mathfrak{F}$ by $\varphi(X) = 1_X:X \to X$ and $\varphi(f) = (f,f)$.

<u>Claim</u>: φ is a full embedding. Clearly φ is an embedding. Now if

$$\begin{array}{ccc} X & \xrightarrow{1_X} & X \\ {\scriptstyle m'}\downarrow & & \downarrow{\scriptstyle m''} \\ Y & \xrightarrow{1_X} & Y \end{array}$$

commutes then $m' = m''$ so $\varphi(m') = (m',m'')$. Thus it is full. From now on then we consider $\mathfrak{S} \subset \mathfrak{F}$.

<u>Theorem 4.1</u>: $\mathfrak{F}$ is a frobenius category with $\mathfrak{S}$ generating the full subcategory of projective-injectives. Furthermore if $T:\mathfrak{S} \to \mathfrak{A}$ is an additive functor into an abelian category $\mathfrak{A}$ then there exist functors $R,M,L:\mathfrak{F} \to \mathfrak{A}$ each extending T, and R is right exact, L left exact and M preserves images.

We shall now prove this theorem keeping in mind that once a lemma is proved its dual is automatically proved (using Spanier-Whitehead duality which clearly extends to $\mathfrak{F}$).

We shall write objects of $\mathfrak{F}$ as $\alpha = (A' \overset{a}{\to} A)$, $\beta = (B' \overset{b}{\to} B)$ and a map $f:\alpha \to \beta$ as

$$\begin{array}{ccc} (A' & \xrightarrow{a} & A) \\ {\scriptstyle f'}\downarrow & & \downarrow{\scriptstyle f''} \\ (B' & \xrightarrow{b} & B) \end{array}$$

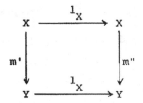

First we shall prove several lemmas which imply that $\mathfrak{F}$ is an abelian category. By Freyd's "recognition lemma" (<u>Freyd</u> [1]) it will suffice to show that kernels, images and cokernels exist: (in categorical language the word kernel is used for both the object and the inclusion map into the domain object--similarly, a cokernel is the object and the projection), that all monomorphisms are kernels and all epimorphisms are cokernels, and finite direct sums exist.

<u>Lemma 4.2</u>: Given maps in $\mathfrak{J}$

$$f = \begin{array}{ccc} (A' & \xrightarrow{a} & A) \\ {\scriptstyle f'}\downarrow & & \downarrow{\scriptstyle f''} \\ (B' & \xrightarrow{b} & B) \end{array} \qquad g = \begin{array}{ccc} (A' & \xrightarrow{a} & A) \\ {\scriptstyle g'}\downarrow & & \downarrow{\scriptstyle g''} \\ (B' & \xrightarrow{b} & B) \end{array}$$

then $f = g$ if either $f' = g'$ or $f'' = g''$.

<u>Proof</u>: Trivial since in either case $f''a = g''a$ and $bf' = bg'$.

<u>Lemma 4.3</u>:

$$f = \begin{array}{ccc} (A' & \longrightarrow & A) \\ {\scriptstyle f'}\downarrow & & \downarrow{\scriptstyle 1} \\ (B' & \longrightarrow & A) \end{array}$$

is a monomorphism in $\mathfrak{J}$.

<u>Proof</u>: If

$$g = \begin{array}{ccc} (X' & \xrightarrow{x} & X) \\ {\scriptstyle g'}\downarrow & & \downarrow{\scriptstyle g''} \\ (A' & \longrightarrow & A) \end{array}$$

is any map and $fg = 0$ then $1 \circ g''x = 0$ so $g''x = 0$ so $g = 0$.

<u>Lemma 4.3 dual</u>:

$$\begin{array}{ccc} (A' & \longrightarrow & A) \\ {\scriptstyle 1}\downarrow & & \downarrow \\ (A' & \longrightarrow & B) \end{array}$$

is an epimorphism in $\mathfrak{F}$.

Given any map

$$\begin{array}{ccc} (A' & \longrightarrow & A) \\ \downarrow & & \downarrow \\ (B' & \longrightarrow & B) \end{array}$$

we may give it an epi-monic factorization:

$$\begin{array}{ccc} (A' & \longrightarrow & A) \\ {\scriptstyle 1}\downarrow & & \downarrow \\ (A' & \longrightarrow & B) \\ \downarrow & & \downarrow{\scriptstyle 1} \\ (B' & \longrightarrow & B) \end{array}$$

Thus the image of the above map is $(A' \to B)$ the composite by either path.

<u>Lemma 4.4</u>: Given

$$f = \begin{array}{ccc} (A' & \xrightarrow{a} & A) \\ {\scriptstyle f'}\downarrow & & \downarrow{\scriptstyle f''} \\ (B' & \xrightarrow{b} & B) \end{array}$$

where $K \overset{k}{\to} A' \xrightarrow{f''a} B$ is a cofibration, then $\ker f$ is

$$h = \begin{array}{ccc} (K & \xrightarrow{ak} & A) \\ {\scriptstyle k}\downarrow & & \downarrow{\scriptstyle 1} \\ (A' & \longrightarrow & A) \end{array}$$

<u>Proof</u>: Clearly $hf = 0$ since $f''ak = 0$. Suppose

$$g = \begin{array}{ccc} (X' & \xrightarrow{x} & X) \\ {\scriptstyle g'}\downarrow & & \downarrow{\scriptstyle g''} \\ (A' & \xrightarrow{a} & A) \end{array}$$

is such that $gf = 0$. Then $X' \xrightarrow{g'} A' \xrightarrow{f''a} B$ is 0 and so there exists $m: X' \to K$ making

$$\begin{array}{ccc} X' & \xrightarrow{g'} & A' \\ & {\scriptstyle m}\searrow & \uparrow{\scriptstyle k} \\ & & K \end{array}$$

commute, and so

$$\begin{array}{ccc} (X' & \longrightarrow & X) \\ {\scriptstyle m}\downarrow & & \downarrow{\scriptstyle g''} \\ (K & \xrightarrow{ak} & A) \\ {\scriptstyle k}\downarrow & & \downarrow{\scriptstyle 1} \\ (A' & \longrightarrow & A) \end{array} \quad = \quad \begin{array}{ccc} (X' & \longrightarrow & X) \\ \downarrow & & \downarrow \\ (A' & \longrightarrow & A) \end{array}$$

Since Lemma 4.3 says that $(k,1)$ is a monomorphism, we get uniqueness up to isomorphism.

Dually we get

<u>Lemma 4.5</u>: Given

$$f = \begin{array}{ccc} (A & \xrightarrow{a} & A) \\ {\scriptstyle f'}\downarrow & & \downarrow{\scriptstyle f''} \\ (B' & \xrightarrow{b} & B) \end{array}$$

where $A' \xrightarrow{f''a} B \xrightarrow{c} C$ is a cofibration, then $\operatorname{cok} f$ is

$$\begin{array}{ccc} (B' & \xrightarrow{b} & B) \\ {\scriptstyle 1}\downarrow & & \downarrow{\scriptstyle c} \\ (B' & \xrightarrow{cb} & C) \end{array}$$

Proposition 4.6: If $f:X \to Y$ is a map in $\mathcal{S}$ then in $\mathcal{F}$ there is an exact sequence

$$0 \to \text{cok } f \to C_f \to \ker Sf \to 0$$

Proof: We have $X \xrightarrow{f} Y \xrightarrow{i} C_f \xrightarrow{\sigma} SX \xrightarrow{Sf} SY$. By Lemma 4.4, $\ker Sf = (C_f \xrightarrow{\sigma} SX)$. Map C_f onto $\ker Sf$ by

$$j = \begin{array}{ccc} (C_f & \xrightarrow{1} & C_f) \\ {\scriptstyle 1}\downarrow & & \downarrow{\scriptstyle \sigma} \\ (C_f & \xrightarrow{\sigma} & SX) \end{array}$$

Then $\ker j = (Y \xrightarrow{i} C_f) = \text{cok } f$.

Lemma 4.7: A map

$$h = \begin{array}{ccc} (A' & \xrightarrow{a} & A) \\ {\scriptstyle h'}\downarrow & & \downarrow{\scriptstyle 1} \\ (B' & \longrightarrow & A) \end{array}$$

is a kernel in $\mathcal{F}$.

Proof: Let $A' \xrightarrow{a} A \xrightarrow{f'} F$ be a cofibration. Then let

$$f = \begin{array}{ccc} (B' & \longrightarrow & A) \\ {\scriptstyle 1}\big\downarrow & & \big\downarrow{\scriptstyle f'} \\ (B' & \longrightarrow & F) \end{array}$$

$fh = 0$ since $f'a = 0$. Suppose

$$g = \begin{array}{ccc} (X' & \longrightarrow & X) \\ {\scriptstyle g'}\big\downarrow & & \big\downarrow{\scriptstyle g''} \\ (B' & \longrightarrow & A) \end{array}$$

is such that $gh = 0$. Then $X' \to A \to F$ is 0 so there is a lifting $X' \to A'$. Let

$$k = \begin{array}{ccc} (X' & \longrightarrow & X) \\ {\scriptstyle k'}\big\downarrow & & \big\downarrow{\scriptstyle k''} \\ (A' & \xrightarrow{a} & A) \end{array} .$$

Then $hk = g$. Thus h is the kernel of f .

<u>Lemma 4.7 dual</u>: A map

$$\begin{array}{ccc} (A' & \longrightarrow & A) \\ {\scriptstyle 1}\big\downarrow & & \big\downarrow \\ (A' & \longrightarrow & B) \end{array}$$

is a cokernel.

Next let

$$f = \begin{array}{ccc} (A' & \xrightarrow{a} & A) \\ {\scriptstyle f'}\big\downarrow & & \big\downarrow{\scriptstyle f''} \\ (B' & \xrightarrow{b} & B) \end{array}$$

be any monomorphism. We wish to show that f is a kernel. Let $K \overset{\sigma}{\to} A' \xrightarrow{f''a} B$ and $K' \overset{\sigma'}{\to} A' \overset{a}{\to} A$ be cofibrations. Since f is a monomorphism, $0 = \ker f = (K \overset{\sigma}{\to} A' \overset{a}{\to} A)$. Thus for some map $\alpha : K \to K'$ we have $\sigma = \sigma'\alpha$. Thus

$$\begin{array}{ccc} K & \xrightarrow{\sigma} & A' \\ {\scriptstyle\alpha}\downarrow & & \downarrow{\scriptstyle 1} \\ K' & \xrightarrow{\sigma'} & A' \end{array}$$

commutes, there is a map $\beta : B \to A$ such that

$$\begin{array}{ccc} A' & \xrightarrow{f''a} & B \\ {\scriptstyle 1}\downarrow & & \downarrow{\scriptstyle \beta} \\ A' & \xrightarrow{a} & A \end{array}$$

commutes. Then

$$\begin{array}{ccc} (A' & \xrightarrow{f''a} & B) \\ {\scriptstyle 1}\downarrow & & \downarrow{\scriptstyle \beta} \\ (A' & \xrightarrow{a} & A) \end{array}$$

is an inverse to

$$g = \begin{array}{ccc} (A' & \xrightarrow{a} & A) \\ {\scriptstyle 1}\downarrow & & \downarrow{\scriptstyle f''} \\ (A' & \xrightarrow{f''a} & B) \end{array}$$

hence they are isomorphisms. Then since

$$k = \begin{array}{ccc} (A' & \xrightarrow{f''a} & B) \\ {\scriptstyle f'}\downarrow & & \downarrow{\scriptstyle 1} \\ (B' & \xrightarrow{b} & B) \end{array}$$

is a kernel by Lemma 4.7, so is $kg = f$.

Thus we have shown that monomorphisms are kernels. Dually epimorphisms are cokernels. Finally setting $(A' \to A) \oplus (B' \to B) = (A' \oplus B' \to A \oplus B)$ we get finite direct sums. Thus the Freyd Category is abelian.

Let us look at the ring End X for some object X of $\mathcal{S}$

Since End $\mathcal{X}$ = End $S^n X$ for all n, we may as well assume X is real. Let $e \in$ End X be an idempotent; i.e. $e = e^2$ in End X. By convention idempotent will mean $\neq 0, 1$.

We would like to prove that $X = Y \vee Y'$ for some space Y and

$$e = \begin{pmatrix} 1_Y & 0 \\ 0 & 0 \end{pmatrix}.$$

Form the space (not in $\mathcal{S}$) $W = \bigvee_{i=1}^{\infty} X$ and map $f: W \to W$ by

$$\begin{pmatrix} 1-e & 0 & 0 & 0 & \\ e & 1-e & 0 & 0 & \\ 0 & e & 1-e & 0 & \\ 0 & 0 & e & 1-e & 0 \\ & & & & \ddots \end{pmatrix}$$

Let $Y = C_f$, $j: W \to Y$, $\sigma: S^{-1}Y \to W$ induced. (Observe that if $e = 0$, $Y = *$; if $e = 1$, $Y = X$.)

Claim: $j = (a,0,0,\ldots)$ where $a(1-e) = 0$.

Proof: Let $j = (a_1, a_2, \ldots): W \to Y$. Then $jf = 0$ so

$$0 = (a_1, a_2, \ldots) \begin{pmatrix} 1-e & & \\ e & 1-e & \\ & e & \\ & & \ddots \end{pmatrix} =$$

$(a_1 - a_1 e + a_2 e,\ a_2 - a_2 e + a_3 e,\ \ldots)$. Thus for all $i \geq 1$ $a_i(1-e) + a_{i+1}e = 0$. Right multiplication by e yields $a_{i+1}e = 0$ for $i \geq 1$. Thus $a_i(1-e) = 0$ for $i \geq 1$. But then for $i \geq 2$ $a_i = a_i e = 0$. Thus $j = (a_1, 0, 0, \ldots)$ where $a_1(1-e) = 0$.

Claim: $\sigma = 0$.

Proof: Let $\sigma = \begin{pmatrix} a_1 \\ a_2 \\ \vdots \end{pmatrix} : S^{-1}Y \to W$. Then

$$0 = f\sigma = \begin{pmatrix} 1-e & & \\ e & 1-e & \\ & e & \ddots \end{pmatrix} \begin{pmatrix} a_1 \\ a_2 \\ \vdots \end{pmatrix} = \begin{pmatrix} a_1(1-e) \\ a_1 e + a_2(1-e) \\ \vdots \end{pmatrix}$$

Thus $a_n e + a_{n+1}(1-e) = 0$ for all $n \geq 1$ and $a_1(1-e) = 0$. Right multiplication by e yields $a_n e = 0$ for all $n \geq 1$ so $a_{n+1}(1-e) = 0$ for all $n \geq 1$ and $a_1(1-e) = 0$. Thus for all $n \geq 1$ $0 = a_n(1-e) = a_n - a_n e = a_n$. Thus $\sigma = 0$.

Now look at the following commutative diagram:

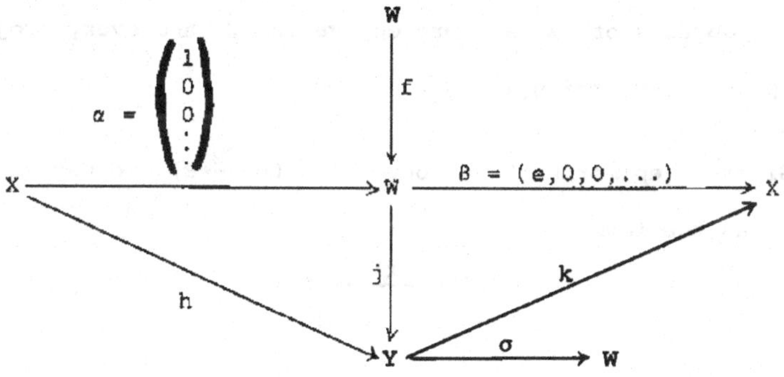

k exists since $\beta f = 0$. $kh = \beta\alpha = e$. Look at $(1-hk): Y \to Y$.

$j^*(1-hk) = (1-hk)j = j - hkj = j - j\alpha kj = j - j\alpha\beta = j(1-\alpha\beta) =$

$$= j \begin{pmatrix} 1-e & 0 & 0 & \cdots & 0 \\ 0 & 1 & 0 & \cdots & 0 \\ 0 & 0 & 1 & \cdots & 0 \\ & & & \ddots & \\ 0 & 0 & 0 & & 1 \end{pmatrix} = (a,0,0,\ldots) \times \begin{pmatrix} 1-e & 0 & \cdots & 0 \\ 0 & 1 & & \\ 0 & 0 & \cdots & 1 \end{pmatrix} = 0 .$$

But since $\sigma^* = 0$, j^* is a monomorphism. Thus $1 - hk = 0$ so $hk = 1$. Thus we have a cofibration $Y' \to X \underset{k}{\overset{h}{\rightleftarrows}} Y$ since $hk = 1$, $X = Y \vee Y'$. $e = \begin{pmatrix} 1_Y & 0 \\ 0 & 0 \end{pmatrix}$.

Conversely if $X = Y \vee Y'$ then $e = \begin{pmatrix} 1_Y & 0 \\ 0 & 0 \end{pmatrix}$ is an idempotent.

Thus we have proved

<u>Theorem 4.8</u>: X is wedge indecomposable if and only if End X has no idempotents.

We use this theorem to prove that $\mathfrak{F}$ is Frobenius with $\mathfrak{S}$ the subcategory of projective-injectives.

<u>Lemma 4.9</u>: Objects of $\mathfrak{S}$ are projective in $\mathfrak{F}$ and every projective is isomorphic to something in $\mathfrak{S}$.

<u>Proof</u>: Given an epimorphism g onto $B = (B \xrightarrow{1_B} B)$ we may as well assume it has the form

$$\begin{array}{ccc} (B & \xrightarrow{f'} & A) \\ {\scriptstyle 1}\downarrow & & \downarrow{\scriptstyle g'} \\ (B & \xrightarrow{1} & B) \end{array} .$$

But then

$$\begin{array}{ccc} (B & \xrightarrow{1} & B) \\ {\scriptstyle 1}\downarrow & & \downarrow{\scriptstyle f'} \\ (B & \xrightarrow{f'} & A) \end{array}$$

splits g . Conversely if $(A \xrightarrow{f} B)$ is projective then the epimorphism

$$\begin{array}{ccc} (A & \xrightarrow{1} & A) \\ {\scriptstyle 1}\downarrow & & \downarrow{\scriptstyle f} \\ (A & \xrightarrow{f} & B) \end{array}$$

is split by some map

$$\begin{array}{ccc} (A & \xrightarrow{f} & B) \\ {\scriptstyle 1}\downarrow & & \downarrow{\scriptstyle g} \\ (A & \xrightarrow{1} & A) \end{array} \quad .$$

This says that $fgf = f$. Thus letting $e = fg{:}B \to B$ and $k = gf{:}A \to A$ we have $e^2 = e$ and $k^2 = k$. By Theorem 4.8, then, $A \cong A' \vee A''$ and $B \cong B' \vee B''$ where $k = \begin{pmatrix} 1_{A'} & 0 \\ 0 & 0 \end{pmatrix}$ and $e = \begin{pmatrix} 1_{B'} & 0 \\ 0 & 0 \end{pmatrix}$. Representing $f = \begin{pmatrix} a & b \\ c & d \end{pmatrix}$ and $g = \begin{pmatrix} \bar{a} & \bar{b} \\ \bar{c} & \bar{d} \end{pmatrix}$, the equations $fk = f$ and $ef = f$ lead to the fact that $b = 0$, $c = 0$, and $d = 0$. Furthermore, $fg = e$ implies $a\bar{a} = 1_{B'}$ and $gf = k$ implies $\bar{a}a = 1_{A'}$. Thus $(A' \xrightarrow{a} B') \cong (A' \xrightarrow{1} A')$. Thus

$$(A \xrightarrow{f} B) \cong (A' \vee A'' \xrightarrow{\begin{pmatrix} a & 0 \\ 0 & 0 \end{pmatrix}} B' \vee B'') \cong (A' \xrightarrow{a} B') \oplus (A'' \xrightarrow{0} B'') \cong$$
$$(A' \xrightarrow{1} A') \oplus 0 \cong (A' \xrightarrow{1} A') .$$ This completes the proof.

Dually everything in $\mathcal{S}$ is $\mathcal{J}$-injective and every $\mathcal{J}$-injective is isomorphic to something in $\mathcal{S}$.

We have enough projectives since for any object $A' \to A$ we have the epimorphism

$$\begin{array}{ccc} (A' & \xrightarrow{1} & A') \\ {\scriptstyle 1}\big\downarrow & & \big\downarrow \\ (A' & \longrightarrow & A) \end{array}$$

and the monomorphism

$$\begin{array}{ccc} (A' & \longrightarrow & A) \\ \big\downarrow & & \big\downarrow{\scriptstyle 1} \\ (A & \xrightarrow{1} & A) \end{array} .$$

Finally given a functor $T: \mathfrak{S} \to \mathfrak{A}$, $\mathfrak{A}$ an abelian category, define $R(A' \to A) = \operatorname{Cok}(T(K) \to T(A'))$

$M(A' \to A) = \operatorname{Im}(TA' \to TA)$

$L(A' \to A) = \ker(TA \to T(SK))$

where $K \to A' \to A$ is a cofibration. (Observe that if $(A' \to A) = (A \xrightarrow{1} A)$ then $K \cong 0$ so $R(A \xrightarrow{1} A) = TA$, $M(A \xrightarrow{1} A) = TA$, $L(A \xrightarrow{1} A) = TA)$. This completes the proof of Theorem 4.1

Observe that a sequence in $\mathfrak{S}$, $X \xrightarrow{f} Y \xrightarrow{g} Z$ is exact in $\mathfrak{F}$ if and only if $(X \xrightarrow{f} Y) \cong (S^{-1}C_g \xrightarrow{\sigma} Y)$. Thus in particular if $X \xrightarrow{f} Y \xrightarrow{g} Z$ is a cofibration, it is exact. (We cannot say "only if" because, for example, $X \vee X' \xrightarrow{(f,0)} Y \xrightarrow{\binom{g}{0}} Z \vee Z'$ is also exact.)

Next we look at some results of Serre. Cf. Hu [1].

Definition: A perfect and weakly complete (pwc) class of abelian groups C is a collection of abelian groups such that

1) $0 \in C$

2) if $A \cong B$ and $B \in C$ then $A \in C$

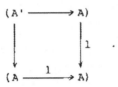

3) if $0 \to A \to B \to C \to 0$ is exact, then $B \in C \Leftrightarrow A, C \in C$

4) $A, B \in C \Rightarrow A \otimes B$, $\mathrm{Tor}(A,B) \in C$

5) $A \in C \Rightarrow \tilde{H}_m(A) \in C$, $m > 0$

<u>Definition</u>: A C-monomorphism is a map $f: A \to B$ such that $\ker f \in C$. A C-epimorphism is a map such that $\mathrm{cok}\, f \in C$, C-isomorphism is the equivalence relation generated by maps which are both C-monomorphisms and C-epimorphisms. Serre proved the following "mod C Hurewicz Theorem."

<u>Theorem 4.10</u>: Let C be pwc and let X be simply connected. If $\pi_i(X) \in C$ for $i < n$ then $h_n: \pi_n(X) \to H_n(X)$ is a C-isomorphism.

<u>Proof</u>: One can show by a fairly easy Serre Spectral Sequence argument that if there is a spectral sequence $H_*(B; H_*(F)) \Rightarrow H_*(E)$ and if $\tilde{H}_*(F)$ or $\tilde{H}_*(B) \in C$ then the other is C-isomorphic to $H_*(E)$. Then the fibration $K(\pi, n-1) \to * \to K(\pi, n)$ yields the fact that if $\tilde{H}_*(K(\pi, n-1)) \in C$ then $\tilde{H}_*(K(\pi, n)) \in C$. But if $\pi \in C$, $\tilde{H}_*(K(\pi, 1)) = \tilde{H}_*(\pi) \in C$, hence $\tilde{H}_*(K(\pi, n)) \in C$ for all n . Now given $\pi_i(X) \in C$ for $i < n$ simply kill off all homotopy groups below dimension n by mapping into $K(\pi_i(X), i)$'s for $i < n$. This will not affect the C -isomorphism class of $H_*(X)$ or $\pi_*(X)$ and then one can apply the usual Hurewicz Theorem.

Examples of such classes: 1) the class of finite abelian groups 2) for a prime p the class of groups with all element of finite order prime to p , $\mathbb{Q}_p$. 3) The class of groups of order one. 4) The class of finitely generated abelian groups , $\mathfrak{FG}$.

Let Q be the rational numbers, then classically one knows that $\tilde{H}_i(Q) = \begin{cases} Q & i = 1 \\ 0 & i \neq 1 \end{cases}$. Then using the Serre Spectral Sequence for $K(Q,1) \to * \to K(Q,2)$ we get easily that $\tilde{H}_i(K(Q,2)) = \begin{cases} Q & i = 2n > 0 \\ 0 & i = 2n+1,\ 0 \end{cases}$ and inductively we find that $\tilde{H}_i(Q,n) = Q$ for $i = n$ for n odd and all multiples of n for n even and 0 otherwise. Thus $H_i(\underline{K}(Q)) = \begin{cases} Q & i = 0 \\ 0 & i \neq 0 \end{cases}$, where $\underline{K}(Q)$ is the Eilenberg-MacLane spectrum of Q. Using the Whitehead Spectral Sequence to be developed in Chapter 5, it is immediate that $H_*(\underline{K}(Z)) \otimes Q = H_*(\underline{K}(Q))$, hence $H_i(\underline{K}(Z))$ is a torsion group for $i > 0$. Since $Z \in \mathfrak{FG}$, $H_i(\underline{K}(Z)) \in \mathfrak{FG}$, thus in fact

<u>Theorem 4.11</u>: $H_i(\underline{K}(Z))$ is finite for $i > 0$.

Thus $\iota : \underline{S} \to \underline{K}(Z)$ induces a $\mathbb{Q}$-isomorphism in homology, hence, giving the obvious meaning to it, is a $\mathbb{Q}$-w.h.e. Thus the map of homology theories $h : \pi_*^s(\ ) = H_*(\ ;\underline{S}) \to H_*(\ ;\underline{K}) = H_*(\ )$ is a $\mathbb{Q}$-isomorphism.

<u>Corollary 4.12</u>: The Hurewicz map $h : \pi_*^s(\ ) \to H_*(\ )$ is a $\mathbb{Q}$-isomorphism.

<u>Corollary 4.13</u>: $\pi_i(\underline{S})$ is finite for $i > 0$.

From now on, by "space" we shall mean any object of $\mathfrak{S}$; as opposed to non-projective objects of $\mathfrak{F}$.

4.2 Problems in Classification

We want to consider means for classifying spaces up to S-type. The first hope, then, would be that one can find all wedge indecomposable spaces and say that all spaces can be written as a unique sum (i.e. wedge) of such spaces. Unfortunately such a program fails in two respects:

A) unique factorization fails--we can find non-isomorphic indecomposable spaces A, B, C, D, E such that $A \vee B \simeq C \vee D \vee E$.

B) cancellation fails--we can find spaces $A \not\simeq B$ such that $A \vee C \simeq B \vee C$, for some space C.

To construct examples, we use the fact (to be calculated in Chapter 5) that there is a map $f: S^7 \to S^0$ of order 15 and that $\{S^8, S^0\}$ has no elements of order 3. This mixing of primes causes problems. We can get the following:

A) $(S^0 \cup_{3f} e^8) \vee (S^0 \cup_{5f} e^8) \simeq S^8 \vee S^0 \vee (S^0 \cup_{8f} e^8)$

B) $S^0 \cup_{3f} e^8 \not\simeq S^0 \cup_{9f} e^8$ but $S^0 \vee (S^0 \cup_{3f} e^8) \simeq S^0 \vee (S^0 \cup_{9f} e^9)$

Both of these are special cases of some general results which follow from <u>algebraic</u> considerations, but we first give geometric proofs of these facts:

Look at the following diagram

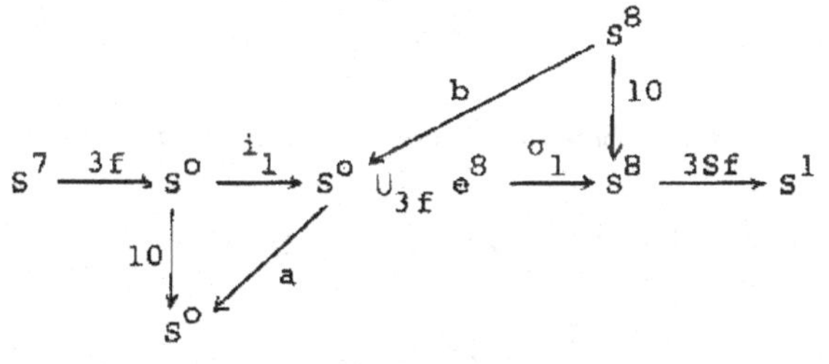

We can find maps a and b making the above diagrams commute because $10\cdot 3f = 0$ and $3Sf\cdot 10 = 0$. Similarly we can find maps a' and b' such that

$$\begin{array}{ccccccccc} & & & & & & S^8 & & \\ & & & & & \overset{b'}{\swarrow} & \downarrow 9 & & \\ S^7 & \xrightarrow{5f} & S^0 & \xrightarrow{i_2} & S^0 \cup_{5f} e^8 & \xrightarrow{\sigma_2} & S^8 & \xrightarrow{5Sf} & S^1 \\ & & 9\downarrow & \underset{a'}{\swarrow} & & & & & \\ & & S^0 & & & & & & \end{array}$$

commutes.

Since $\{S^8, S^0\}$ is finite and has no elements of order 3, we may divide by powers of 3. Let $k = \frac{1}{9}(ab + a'b')$. Thus $ab + a'b' = 9k = a'i_2k$. Then

$$\begin{pmatrix} a & -a' \\ \sigma_1 & \sigma_2 \end{pmatrix}\begin{pmatrix} i_1 & b \\ i_2 & -b'+i_2k \end{pmatrix} = \begin{pmatrix} ai_1-a'i_2 & ab+a'b'-a'i_2k \\ \sigma_1 i_1+\sigma_2 i_2 & \sigma_1 b-\sigma_2 b'+\sigma_2 i_2 k \end{pmatrix} =$$

$$\begin{pmatrix} 1 & 0 \\ 0 & 1 \end{pmatrix}.$$

That is

$$S^0 \vee S^8 \xrightarrow{\begin{pmatrix} i_1 & b \\ i_2 & -b'+i_2k \end{pmatrix}} (S^0 \cup_{3f} e^8) \vee (S^0 \cup_{5f} e^8) \xrightarrow{\begin{pmatrix} a & -a' \\ \sigma_1 & \sigma_2 \end{pmatrix}} S^0 \vee S^8$$

$$= \text{identity}.$$

Thus $S^0 \vee S^8$ is a summand of the middle space and hence

$$(S^0 \cup_{3f} e^8) \vee (S^0 \cup_{5f} e^8) \simeq S^0 \vee S^8 \vee X$$

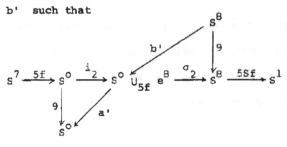

for some space X . By taking homology we see that $X = S^0 \cup e^8$ although we do not have immediately that the attaching map may be taken as $8f$. This will follow from Theorem 4.25. This gives us bad example A.

For B we look at

$$\begin{array}{ccccc}
S^8 & \xrightarrow{\begin{pmatrix}3f\\0\end{pmatrix}} & S^0 \vee S^0 & \longrightarrow & (S^0 \cup_{3f} e^8) \vee S^0 \\
\downarrow{\scriptstyle 1} & & \downarrow{\scriptstyle \begin{pmatrix}3 & 2\\5 & 3\end{pmatrix}} & & \downarrow{\scriptstyle g} \\
S^8 & \xrightarrow{\begin{pmatrix}9f\\0\end{pmatrix}} & S^0 \vee S^0 & \longrightarrow & (S^0 \cup_{9f} e^8) \vee S^0
\end{array}$$

The left hand square commutes, so there is a map g induced. Since 1 and $\begin{pmatrix}3 & 2\\5 & 3\end{pmatrix}$ are equivalences (the latter since its determinant is -1), so is g . (Use mapping cone property 5.) On the other hand, since $9f \neq \pm 3f$, the following theorem implies $S^0 \cup_{3f} e^8 \not\simeq S^0 \cup_{9f} e^8$; hence we have bad example B.

<u>Theorem 4.14</u>: If $f,g: S^{n-1} \to S^m$, then $S^m \cup_f e^n \simeq S^m \cup_g e^n$ if and only if $f = \pm g$.

<u>Proof</u>: In order that f or g be non-trivial, it is necessary to assume that $n-1 \geq m$. By taking homology groups, the result is easily seen to be true for $n - 1 = m$ since $H_n(S^m \cup_f e^{n+1}) \simeq Z_d$ where d is the absolute value of the degree of f . So we assume $n \geq m + 2$. Then look at

$$\begin{array}{ccccccc}
S^{n-1} & \xrightarrow{f} & S^m & \xrightarrow{i_1} & S^m \cup_f e^n & \xrightarrow{\sigma_1} & S^n \\
\downarrow{\scriptstyle a} & & \downarrow{\scriptstyle b} & & \downarrow{\scriptstyle c} & & \downarrow{\scriptstyle Sa} \\
S^{n-1} & \xrightarrow{g} & S^m & \xrightarrow{i_2} & S^m \cup_g e^n & \xrightarrow{\sigma_2} & S^n
\end{array}$$

If $f = \pm g$, then taking $a = 1$, $b = \pm 1$ yields a map c which is an equivalence since a and b are. Conversely given an equivalence c , $\sigma_2 c i_1 : S^m \to S^n$ is 0 since $m < n$. Thus there exists a map b making the diagram commute, hence a map Sa is induced, thus we can desuspend to get a . Since c is an equivalence, using Property 5, we get that $S^{n-1} \cup_a e^n \simeq S^m \cup_b e^{m+1}$. Since $n-1 \neq m$ $H_m(S^{n-1} \cup_a e^n) = 0$. Thus $H_m(S^m \cup_b e^{m+1}) = 0$ so b is of degree ± 1 so $f = \pm g$.

Because of these difficulties, the closest we can hope to come in getting a good description of equivalence classes of spaces will necessitate weakening the equivalence: write $X \equiv Y$, X congruent to Y , if and only if for some space Z , $X \vee Z \simeq Y \vee Z$. Thus $\equiv$ is a cancellation relation since clearly $X \vee Z \equiv Y \vee Z$ implies $X \equiv Y$. We still, however, have bad example A to worry about. Even the congruence classes are not freely generated by the indecomposables. This necessitates passing from the cancellation semigroup of congruence classes to the group, $G = K_o(\mathfrak{S})$: the generators of the abelian group G are congruence classes of spaces and we have the relation $[X \vee Y] = [X] + [Y]$. G is the Grothendieck group of $\mathfrak{S}$. Cf. Freyd [3]. We then get the following very nice result:

Theorem 4.15: G is free.

Proving this and describing the generators of G will be the main goal of the remainder of this chapter. Let us see what this says: in the semigroup of congruence classes we have (looking at bad example A) $[S^o \cup_{5f} e^8] + [S^o \cup_{3f} e^8] = [S^o] + [S^8] + [S^o \cup_{8f} e^8]$ where all five are indecomposable. But passing to G , we will not let $S^o \cup_{8f} e^8$

be a generator. The other four will -- not only are they indecomposable, but their attaching maps are of prime power order. We do not allow mixing of primes: 5f is of order 3, 3f of order 5, while 8f is of order $3\cdot 5$. Thus in G we will have $[S^o \cup_{8f} e^8]$ written uniquely as $[S^o \cup_{5f} e^8] + [S^o \cup_{3f} e^8] - [S^o] - [S^8]$.

In order to study congruence, we need to know when we can be sure of cancellation. Recall

<u>Definition</u>: A ring R is local if and only if the set of non-units forms an ideal.

Remarks:

1) This is sometimes called quasi-local since we did not specify Noetherian. (In fact we will only deal with Noetherian although we shall prove some results that don't require it.)

2) The requirement that the non-units form an ideal is satisfied if and only if the sum of non-units is always a non-unit (since the product of a non-unit and anything is a non-unit).

3) If $e \in R$ is an idempotent then e , $(1-e)$ are non-units, but $e + (1-e) = 1$; hence R is not local. If $uu' = 1$ then $(u'u)^2 = (u'u)$ so $u'u$ is $0,1$ or an idempotent; $u'u = 0$ implies $0 = u\cdot(u'u) = (uu')u = u$ a contradiction. Thus if a ring has no idempotents then any right or left unit is a two-sided unit (since $uu' = 1 \Rightarrow u'u = 1$) . Thus the definition of local need not specify further in using the word "unit".

The following result leads to our interest in local rings:

<u>Theorem 4.16</u>: Let $\mathcal{C}$ be an additive category. Let A be an object with End A a local ring. Then

1) if A is a retract of $B \oplus C$ then A is a retract of B or C .

2) $A \oplus B_1 \simeq A \oplus B_2 \Rightarrow B_1 \simeq B_2$.

Proof:

1)

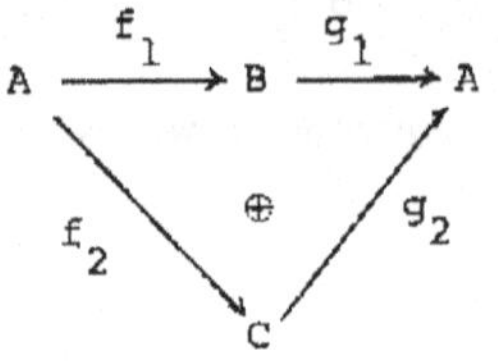

$g_1f_1 + g_2f_2 = 1_A$ thus g_1f_1 or g_2f_2 is a unit (since the sum of non-units is a non-unit). Say $h(g_1f_1) = 1_A$, $h:A \to A$; then $A \xrightarrow{f_1} B \xrightarrow{hg_1} A = 1_A$ so A is a retract of B . (Similarly if g_2f_2 is a unit A is a retract of C).

2) If $A \oplus B_1 \simeq A \oplus B_2$ we have the isomorphism

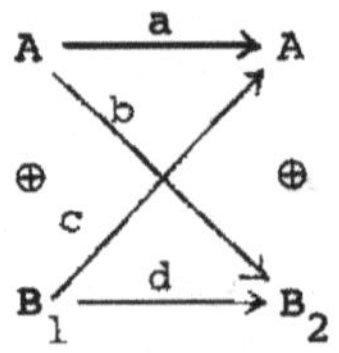

let a',b',c',d' represent the inverse of this isomorphism.

case i) a is a unit

case ii) if not, then

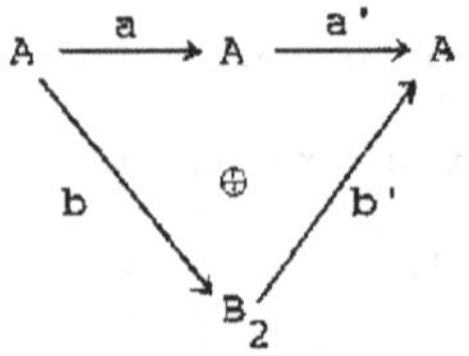

is the identity so A is a retract of B_2 and b'b is a unit. $B_2 \simeq A \oplus B'$. Assume $d:B_1 \to B_2 \simeq A \oplus B'$ is given by $\begin{pmatrix} d_1 \\ d_2 \end{pmatrix}$, b given by $\begin{pmatrix} b_1 \\ b_2 \end{pmatrix}$, b_1 a unit. So we get

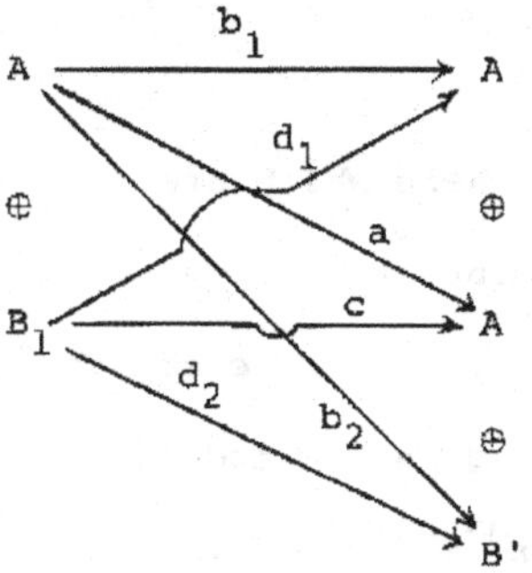

with b_1 a unit.

So we might as well assume that in the original diagram a is a unit. Then followed by

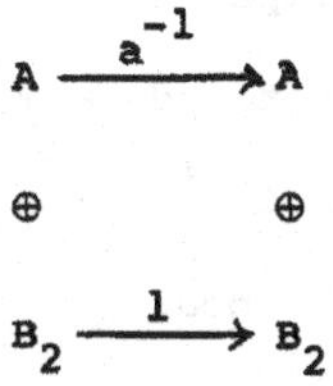

we may as well assume the original diagram is

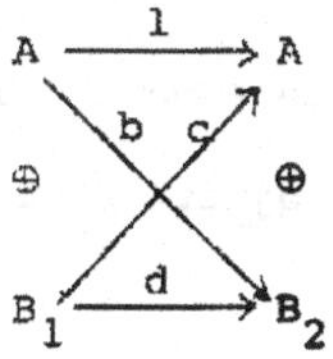

an isomorphism. If b', c' , and d' are as above, then $(d' - b'c')$ is a two-sided inverse to d .

So d is an isomorphism.

A particular case that we meet frequently is the following:

__Proposition 4.17:__ If R is a finite ring without idempotents, then it is local.

Proof: Two points arise here which are very useful facts about a general ring and an element x .

Fact 1: If $\{x,x^2,\ldots\}$ is a finite set then for some m , $x^m = (x^m)^2$: choose integers n and j such that $x^{j+n} = x^j$. Then $x^{nj+n} = x^{nj}$ so $x^{nj}x^{nj} = x^{nj}$.

Fact 2: If x is nilpotent, then $1 - x$ is a unit: if $x^{n+1} = 0$, then $(1+x+x^2+\ldots+x^n)$ is a 2-sided inverse to $1 - x$.

By Fact 1, in a finite ring R without idempotents, every element x is a unit or is nilpotent: $x^m = 0$ or 1 . Thus if a is a non-unit, and u is a unit and $a + b = u$, then au^{-1} is a non-unit, hence nilpotent, so $bu^{-1} = 1 - au^{-1}$ is a unit, by Fact 2, so b is a unit. Thus R is local.

Definition: In an abelian category $\mathfrak{C}$, an object X is called torsion if End X is finite. The order of 1_X is $\exp X$, the exponent of X . In $\mathfrak{F}$, since endomorphism rings are finitely generated abelian groups, an object X is torsion if and only if 1_X is of finite order.

Theorem 4.18: If A is a torsion object of an abelian category $\mathfrak{C}$ and $A \oplus B \cong A \oplus C$, then $B \cong C$.

Proof: Write $A \cong A_1 \oplus \ldots \oplus A_n$ where each A_i is indecomposable. Then End A_i is finite without idempotents, hence local. Then applying Theorem 4.16 n times, $A_1 \oplus \ldots \oplus A_n \oplus B \cong A_1 \oplus \ldots \oplus A_n \oplus C$ implies $B \cong C$.

Lemma 4.19: If X is a finite CW complex and C is a space such that $\pi_*(C)$ is finite in each degree, then $\{X,C\}$ is finite. Thus if C is a finite CW complex and $\pi_*(C)$ finite then C is torsion.

Proof: By induction on the number of cells of X: If $X = S^n$ then $\{S^n,C\} = \pi_n(C)$ is finite. If $\{X,C\}$ is finite when X has $(r-1)$-cells then let $Y = \text{cone}(S^n \to X)$ have r cells. Then

$$\{S^{n+1},C\} \to \{Y,C\} \to \{X,C\}$$

is exact and since $\{X,C\}$ and $\{S^{n+1},C\}$ are finite so is $\{Y,C\}$.

Theorem 4.20: If $0 \to A \overset{f}{\to} B \overset{g}{\to} C \to 0$ is exact, then B is torsion if and only if A and C are.

Proof: $\Rightarrow$: If $m\,1_B = 0$ then $f(m\,1_A) = mf(1_A) = 0$ so $m\,1_A = 0$. $(m\,1_C)g = 1_C(mg) = 0$ so $m1_C = 0$. $\Leftarrow$ Assume $m\,1_A = 0$, $n\,1_C = 0$. Then $g(n\,1_B) = 0$ so $n\,1_B = fh$, $h:B \to A$. Then $m\,n\,1_B = m\,f\,h = 0$.

Schanuel's Lemma 4.21: Let $\mathcal{C}$ be an abelian category with $P_1,\ldots,P_r$ and $Q_1,\ldots,Q_r$ projectives. If

$$0 \to K \to P_r \to P_{r-1} \to \cdots \to P_1 \to M \to 0$$

$$0 \to L \to Q_r \to \cdots \to Q_1 \to M \to 0$$

are exact, then $K \oplus Q_r \oplus P_{r-1} \oplus Q_{r-2} \oplus P_{r-3} \oplus \cdots \cong L \oplus P_r \oplus Q_{r-1} \oplus P_{r-2} \oplus \cdots$.

Proof: Let P and Q be projective. Given the exact sequences

$$0 \to K \xrightarrow{f'} P \xrightarrow{f} M \to 0$$

and

$$0 \to L \xrightarrow{g'} Q \xrightarrow{g} M \to 0 ,$$

let D be the kernel of the map $(f,-g):P \oplus Q \to M$. $\begin{pmatrix}0\\g'\end{pmatrix}:L \to P \oplus Q$ lands in Ker$(f,-g)$ since the composite is 0 . Thus we get a map $h:L \to D$. Since $\begin{pmatrix}0\\g'\end{pmatrix}$ is monic, so is h . But cok $h \cong P$ because the kernel of $D \to P \oplus Q \to P$ is isomorphic to $h:L \to D$. Thus $D \cong P \oplus L$. Similarly $D \cong Q \oplus K$. Thus the lemma is proved for $r = 1$.

Inductively assume the lemma holds for $r - 1 \geq 1$. Then given the exact sequences

$$0 \to K \to P_r \xrightarrow{f_r} P_{r-1} \to \cdots \to P_1 \to M \to 0$$

and

$$0 \to L \to Q_r \xrightarrow{g_r} Q_{r-1} \to \cdots \to Q_1 \to M \to 0$$

we get the exact sequences

$$0 \to \text{im } f_r \to P_{r-1} \to \cdots \to P_1 \to M \to 0$$

and

$$0 \to \text{im } g_r \to Q_{r-1} \to \cdots \to Q_1 \to M \to 0$$

so that by induction

$$\text{im } f_r \oplus Q_{r-1} \oplus P_{r-2} \oplus \cdots \cong \text{im } g_r \oplus P_{r-1} \oplus Q_{r-2} \oplus \cdots$$

Now $0 \to K \to P_r \to \text{im } f_r \to 0$ and $0 \to L \to Q_r \to \text{im } g_r \to 0$ are exact thus

$$0 \to K \to P_r \oplus (Q_{r-1} \oplus P_{r-2} \oplus \ldots) \to \text{im } f_r \oplus (Q_{r-1} \oplus P_{r-2} \oplus \ldots) \to 0$$

and

$$0 \to L \to Q_r \oplus (P_{r-1} \oplus Q_{r-2} \oplus \ldots) \to \text{im } g_r \oplus (P_{r-1} \oplus Q_{r-2} \oplus \ldots) \to 0$$

are exact. But the third terms in the two sequences are isomorphic. Thus by the case $r = 1$ we get

$$K \oplus Q_r \oplus P_{r-1} \oplus Q_{r-2} \oplus \ldots \cong L \oplus P_r \oplus Q_{r-1} \oplus P_{r-2} \oplus \ldots .$$

We shall frequently use the following form of Schanuel's Lemma.

Special Schanuel Lemma 4.22: Let $\mathcal{G}$ be an abelian category with $P_1,\ldots,P_r$ and $Q_1,\ldots,Q_r$ projective. If

$$0 \to K \to P_r \to \ldots \to P_1 \to M \to 0$$

and

$$0 \to K \to Q_r \to \ldots \to P_1 \to M \to 0$$

are exact where either K or M is torsion, then

$$P_r \oplus Q_{r-1} \oplus \ldots \cong Q_r \oplus P_{r-1} \oplus \ldots .$$

Proof: If K is torsion, then we cancel it from the equation $K \oplus P_r \oplus Q_{r-1} \oplus \ldots \cong K \oplus Q_r \oplus P_{r-1} \oplus \ldots$ that we get from the previous. On the other hand there is a dual version of Lemma 4.21 which gives the same conclusion when all the arrows in the hypothesis are reversed. (We needn't prove it here. In $\mathfrak{F}$ it follows automatically by duality.) Thus if M is torsion we get the same conclusion.

We can now prove some decomposition theorems which, in particular, give an alternate proof of bad example A.

<u>Theorem 4.23</u>: Given $f_i : X_i \to Z$ of order m_i, $i = 1,2$ where $(m_1, m_2) = 1$, form $(f_1, f_2) : X_1 \vee X_2 \to Z$. Then $C_{f_1} \vee C_{f_2} \simeq C_{(f_1,f_2)} \vee Z$.

<u>Proof:</u> By assuming that $X_1 \vee X_2 \subset Z$ we see that $Z/X_1 \vee X_2 \cong (Z/X_1)/X_2$. This says that in the general case the cone of the composite $X_2 \xrightarrow{f_2} Z \xrightarrow{\sigma_1} C_{f_1}$ is $C_{(f_1,f_2)}$. f_2 is the composite $X_2 \overset{\alpha}{\to} \text{im } f_2 \overset{\beta}{\to} Z$. Now β is a monomorphism so $\ker \sigma_1\beta \cong \beta \ker \sigma_1\beta \subset \ker \sigma_1 = \text{im } f_1$. But choosing s_1 and s_2 such that $s_1m_1 + s_2m_2 = 1$ we see that $s_1m_1 = 0$ on $\text{im } f_1$ and $s_1m_1 = 1$ on $\text{im } f_2$, hence s_1m_1 is both 0 and 1 on $\ker \sigma_1\beta$, hence $\ker \sigma_1\beta = 0$. Thus $\sigma_1\beta$ is a monomorphism, thus $\text{im } \sigma_1 f_2 = \text{im } \sigma_1\beta\alpha \cong \text{im } \alpha = \text{im } f_2$ since α is onto. Also $\ker \sigma_1 f_2 = \ker \sigma_1\beta\alpha = \ker \alpha = \ker \beta\alpha$ since β and $\sigma_1\beta$ are monomorphisms. Then since $0 \to \text{im } \sigma_1 f_2 \to C_{f_1} \to C_{(f_1,f_2)} \to S \ker \sigma_1 f_2 \to 0$ is exact, there is an exact sequence

$$0 \to \text{im } f_2 \to C_{f_1} \to C_{(f_1,f_2)} \to \ker Sf_2 \to 0$$

On the other hand, there is an exact sequence

$$0 \to \text{im } f_2 \to Z \to C_{f_2} \to \ker Sf_2 \to 0 .$$

Then by the special Schanuel Lemma, we get that $Z \vee C_{(f_1,f_2)} \simeq C_{f_1} \vee C_{f_2}$.

By induction we extend this to

<u>Corollary 4.24</u>: If $f_i : X_i \to Z$ is of order m_i with the m_i pairwise relatively prime, $i = 1, \ldots, n$, then $\bigvee_{i=1}^{m} C_{f_i} \simeq C_{(f_1, f_2, \ldots, f_n)} \vee \bigvee_{1}^{n-1} Z$.

We observe the special case of $X_1 = X_2$.

<u>Theorem 4.25</u>: If $f_1, f_2 : X \to Z$ are of orders m_1, m_2 with $(m_1, m_2) = 1$, then $C_{(f_1, f_2)} \simeq SX \vee C_{f_1+f_2}$. Thus $C_{f_1} \vee C_{f_2} \simeq SX \vee C_{f_1+f_2} \vee Z$.

<u>Proof</u>: The commutative diagram

$$\begin{array}{ccc} X & \xrightarrow{f_1 + f_2} & Z \\ {\scriptstyle \binom{1}{1}} \big\downarrow & & \big\| \\ X \vee X & \xrightarrow{(f_1, f_2)} & Z \end{array}$$

yields a map $C_{f_1+f_2} \to C_{(f_1, f_2)}$ with cofibre $SC_{\binom{1}{1}}$ by Property 5). But $C_{\binom{1}{1}} \simeq X$. If $s_1 m_1 + s_2 m_2 = 1$, then the commutative diagram

$$\begin{array}{ccc} X \vee X & \xrightarrow{(f_1, f_2)} & Z \\ {\scriptstyle \binom{s_2 m_2}{s_1 m_1}} \big\downarrow & & \big\| \\ X & \xrightarrow{f_1 + f_2} & Z \end{array}$$

yields a map $C_{(f_1,f_2)} \to C_{f_1+f_2}$ which clearly splits $C_{f_1+f_2} \to C_{(f_1,f_2)}$. Thus $C_{(f_1,f_2)} \simeq SX \vee C_{f_1+f_2}$. The second statement follows from Theorem 4.23.

Now bad example A follows: since $3f$ and $5f: S^7 \to S^0$ are of relatively prime orders, $C_{3f} \vee C_{5f} \simeq S^8 \vee S^0 \vee C_{8f}$.

We can also use the special Schanuel Lemma to give an algebraic proof to bad example B. We wish to show that $S^0 \vee (S^0 \cup_{3f} e^8) \simeq S^0 \vee (S^0 \cup_{9f} e^8)$. Since $3f$ is of order 5, $\operatorname{im} 3f$ is also, thus $3\,1_{\operatorname{im} 3f}$ is an isomorphism. Then by the commutativity of

$$\begin{array}{ccc} S^7 & \longrightarrow & \operatorname{im} 3f \\ {\scriptstyle 9f}\downarrow & & \wr\downarrow {\scriptstyle 3 1_{\operatorname{im} 3f}} \\ S^0 & \longleftarrow & \operatorname{im} 3f \end{array}$$

we see that $\operatorname{im} 9f \simeq \operatorname{im} 3f$ and $\ker 3f \simeq \ker 9f$. Thus there are exact sequences

$$0 \to \operatorname{im} 3f \to S^0 \to S^0 \cup_{9f} e^8 \to S \ker 3f \to 0$$

$$0 \to \operatorname{im} 3f \to S^0 \to S^0 \cup_{3f} e^8 \to S \ker 3f \to 0$$

hence $S^0 \vee (S^0 \cup_{3f} e^8) \simeq S^0 \vee (S^0 \cup_{9f} e^8)$. But by Theorem 4.14, $S^0 \cup_{3f} e^8 \not\simeq S^0 \cup_{9f} e^8$.

4.3 Localization and Cancellation

We are going to look at certain "localizations" of the category $\mathcal{S}$. In fact they can be defined for any abelian category $\mathcal{G}$. Let d be an integer or a set of primes. If d is an integer and p is prime then $p|d$ take on its usual meaning, if d is a set of primes then $p|d$ means $p \in d$. Then we form the principal ideal domain $\mathbb{Q}_d = \{ \frac{r}{s}$ rational $|$ prime $p|s \Rightarrow p|d\}$. Thus $\mathbb{Q}_o = \mathbb{Q}$, the rational numbers, and $\mathbb{Q}_1 = Z$, the integers.

Then let $\mathcal{G}^d$ be the category with the same objects as $\mathcal{G}$ and with $\mathcal{G}^d(X,Y) = \mathcal{G}(X,Y) \otimes \mathbb{Q}_d$. Then a morphism $g \in \mathcal{G}^d(X,Y)$ may be expressed as $\frac{f}{s}$ where $f \in \mathcal{G}(X,Y)$ and the primes of s divide d. Thus the primes dividing d become invertible in $\mathcal{G}^d$.

We write $X \simeq_d Y$ if X and Y are isomorphic in $\mathcal{G}^d$. We read this as X and Y are equivalent mod d. If $X \simeq_d Y$, then there are $\mathcal{G}$-morphisms $f:X \to Y$ and $g:Y \to X$ with $gf = m\,1_X$ and $fg = m\,1_Y$ where the primes of m divide d.

Observe that $\mathcal{G}^1 = \mathcal{G}$. $\mathcal{G}^o$ is $\mathcal{G}$ with all torsion killed, and all non-zero integers are units. In general $\mathcal{G}^d$ is "in between": if $p|d$, p-torsion is killed and p is a unit.

Later on we will consider another localization $\mathcal{G}_p$: it is $\mathcal{G}^d$ where d is the set of all primes <u>except</u> p. This will be discussed in detail later.

Observe that $X \simeq_d 0$ if and only if $0 = 1_X \otimes 1 \in \mathcal{G}(X,X) \otimes \mathbb{Q}_d$: i.e. if and only if for some S with $S\,1_X = 0$ we have $1_X \otimes 1 = S{\cdot}1_X \otimes \frac{1}{S} = 0$ making sense; hence $X \simeq_d 0$ if and only if the primes of $\exp(X)$ divide d. Thus $X \simeq_d Y$ if and only if there is some

$f:X \to Y$ with $\ker f$, $\operatorname{cok} f \simeq_d 0$. This is why $X \simeq_d Y$ if and only if there are maps $f:X \to Y$ and $g:Y \to X$ with $fg = m\,1_Y$, $gf = m\,1_X$ where the primes of m divide d.

We can apply Schanuel's Lemma to the category $\mathcal{C}^d$ to get the following mod d version in $\mathcal{C}$:

<u>Schanuel's Lemma mod d 4.21d</u>: Let $\mathcal{C}$ be an abelian category with $P_1,\ldots,P_r$ and $Q_1,\ldots,Q_r$ projective. If

$$0 \to K \to P_r \to P_{r-1} \to \cdots \to P_1 \to M \to 0$$

and

$$0 \to L \to Q_r \to Q_{r-1} \to \cdots \to Q_1 \to M' \to 0$$

are exact and M and M' are equivalent mod d then

$$K \oplus Q_r \oplus P_{r-1} \oplus \cdots \quad \text{and} \quad L \oplus P_r \oplus Q_{r-1} \oplus \cdots$$

are equivalent mod d.

At the same time observe that any finite ring is the ring direct sum of its p-primary components:

$$R = \oplus R_p \quad \text{and} \quad R_p \otimes Q_d = \begin{cases} 0 & \text{if } p \mid d \\ R_p & \text{if } p \nmid d \end{cases}$$

so we can take a torsion object of $\mathfrak{F}$ forget about the p-primary pieces of it for $p \mid d$ and then go through the proof of Theorem 4.18 in $\mathfrak{F}^d$, this time coming up with the mod d version:

<u>Theorem 4.18d</u>: If A is a torsion object of $\mathfrak{F}$ and $A \oplus B \simeq_d A \oplus C$

then $B \simeq_d C$.

Given spaces X and Y we may ask whether or not $X \equiv Y$. Assuming $H_*(X) \simeq H_*(Y)$ and $\pi_*(X) \simeq \pi_*(Y)$ we may proceed to look for spaces A such that $A \vee X \simeq A \vee Y$. If no such A is to be found, the process may continue indefinitely. Our purpose now is to provide a relatively simple "test space." Given X we will find a certain wedge or "bouquet" of spheres B such that $X \equiv Y$ if and only if $X \vee B \simeq Y \vee B$.

Recall that the n-th Betti-number of a space X of finite type is the integer b_n such that $H_n(X) = Z^{b_n} \oplus$ finite group. (From Theorem 4.12 $\pi_n^s(X) \cong Z^{b_n} \oplus$ finite group for the same b_n .) For any X define $B_X = \bigvee_n (\bigvee_1^{b_n} S^n)$. B is a map of the objects of $\mathcal{S}$. B extends to C and to G .

<u>Theorem 4.26</u>: $X \equiv Y$ if and only if $X \vee B_X \simeq Y \vee B_X$.

We shall need some lemmas before we can get to the proof. We first observe that the theorem is symmetric because $X \equiv Y$ implies $H_*(X) \simeq H_*(Y)$ so that $B_X = B_Y$.

We have the functors

$$\pi_*: \mathcal{F} \to \mathcal{G}^Z \ , \quad F: \mathcal{G}^Z \to F\mathcal{G}^Z$$

where $\mathcal{G}^Z$ is the category of Z-graded abelian groups finitely generated in each degree and $F\mathcal{G}^Z$ is the category of Z-graded free abelian groups. $\pi_n = \mathcal{F}(S^n, \ )$. F kills torsion.

Given $X \in \mathcal{S}$, we can choose generators $\varphi_i \in \pi_{n_i}(X)$ which generate $F\pi_*(X)$. Then $B_X = \bigvee S^{n_i}$ and $\varphi: B_X \to X$, $\varphi = (\varphi_1, \varphi_2, \ldots)$, induces an $F\pi_*$-isomorphism.

<u>Lemma 4.27</u>: Given $\varphi_1, \varphi_2 : B \to X$ inducing $F\pi_*$-isomorphisms, then for sufficiently divisible integers t , $\ker(t\varphi_1) \cong \ker(t\varphi_2)$ and $\operatorname{cok}(t\varphi_1) \cong \operatorname{cok}(t\varphi_2)$.

<u>Proof</u>: Let $\theta = (F\pi_*\varphi_2)^{-1}(F\pi_*\varphi_1) : F\pi_*B \cong F\pi_*B$

Since $F\pi_*B$ is just a finite number of Z's , there is a finite matrix A representing θ and then a map $g: B \to B$ represented by A (where integers stand for maps of that degree). Then $F\pi_*(g) = \theta$. So

$$\begin{array}{ccc} B & \xrightarrow{\varphi_1} & X \\ g\downarrow & & \Vert \\ B & \xrightarrow[\varphi_2]{} & X \end{array}$$

is an $F\pi_*$ commutative diagram. g is a homotopy equivalence since it induces a homology isomorphism $(F\pi_*(B) \cong H_*(B))$.

Thus the diagram differs from being commutative only by the torsion in $[B,X]$, which is finitely generated. So choose t sufficiently divisible that $t[B,X]$ has no torsion. Then

$$\begin{array}{ccc} B & \xrightarrow{t\varphi_1} & X \\ g\downarrow & & \downarrow 1_X \\ B & \xrightarrow[t\varphi_2]{} & X \end{array}$$

commutes. g and 1_X are isomorphisms, hence $\ker(t\varphi_1) \cong \ker(t\varphi_2)$

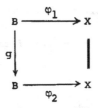

and $\mathrm{cok}(t\varphi_1) \cong \mathrm{cok}(t\varphi_2)$.

Lemma 4.28: If $\varphi : X \to Y$ induces a rational isomorphism (i.e. of homotopy or homology), then $\ker \varphi$ and $\mathrm{cok}\ \varphi$ are torsion objects.

Proof: We have the exact sequence

$$0 \to \mathrm{cok}\ \varphi \to C_\varphi \to S \ker \varphi \to 0$$

and we know that $\exp(S \ker \varphi) = \exp(\ker \varphi)$. Thus from Theorem 4.22 it suffices to show that C_φ is a torsion object. But if φ is a rational isomorphism, the long exact homotopy sequence yields $\pi_*(C_\varphi)$ finite in each degree and then Lemma 4.21 says that C_φ is torsion.

We can now prove Theorem 4.26.

Let Z be such that $X \vee Z \cong Y \vee Z$. Let $B = B_X = B_Y$. Let $\varphi_1 : B \to X$, $\varphi_2 : B \to Y$ and $\varphi : B' \to Z$ be $F\pi_*$-isomorphisms, where $B' = B_Z$. Then $\varphi_1 \oplus \varphi : B \vee B' \to X \vee Z$, $\varphi_2 \oplus \varphi : B \vee B' \to Y \vee Z \cong X \vee Z$ are $F\pi_*$-isomorphisms. So for some $t > 0$

$$\begin{array}{ccc} \ker(t(\varphi_1 \oplus \varphi)) \cong \ker(t(\varphi_2 \oplus \varphi)), & & \mathrm{cok}(t(\varphi_1 \oplus \varphi)) \cong \mathrm{cok}(t(\varphi_2 \oplus \varphi)) \\ \wr\| \qquad\qquad \wr\| & & \\ \ker t\varphi_1 \oplus \ker t\varphi \quad \ker t\varphi_2 \oplus \ker t\varphi & & \end{array}$$

$\ker t\varphi$ is a torsion object by Lemma 4.28. Thus from Corollary 4.18 we get $\ker t\varphi_1 \cong \ker t\varphi_2$. Similarly $\mathrm{cok}\ t\varphi_1 \cong \mathrm{cok}\ t\varphi_2$. Then we have the exact sequences

$$0 \to \ker t\varphi_1 \to B \xrightarrow{t\varphi_1} X \to \operatorname{cok} t\varphi_1 \to 0$$

$$0 \to \ker t\varphi_2 \to B \xrightarrow{t\varphi_2} Y \to \operatorname{cok} t\varphi_2 \to 0$$

so by the special Schanuel Lemma we get $B \oplus X \cong B \oplus Y$; i.e. $B \vee X \cong B \vee Y$.

In the case of a torsion space X , $B_X = 0$ so we get

Corollary 4.29: If X is torsion and $X \equiv Y$ then $X \cong Y$.

The following is an example of a theorem which is more or less "obvious," but which is most efficiently proved algebraically.

Proposition 4.30: In $\mathcal{S}^d$ a retract of a wedge of spheres is a wedge of spheres.

Proof: First we have to make the observation that $\mathbb{Q}_d$ is a principal ideal domain, PID. Hence if M is a free $\mathbb{Q}_d$-module or rank n and $e:M \to M$ is an idempotent then $M = e(M) \oplus (1-e)M$ so $e(M)$ is a free-module of rank m , $(1-e)M$ a free module of rank $(n-m)$. Then we can find a basis for M such that the matrix for e is diagonal with the first m places 1's and zeroes everywhere else. In particular if e is any idempotent $n \times n$ matrix, it is similar to one of the above form.

Then let W be a wedge of spheres, $W = R \vee S$. $e \in \mathcal{S}^d(W,W)$ an idempotent corresponding to $\begin{pmatrix} 1_R & 0 \\ 0 & 0 \end{pmatrix}$. Assume that

$W = \bigvee_{i=1,\dots,n} S^m$ a wedge of spheres all of the same dimension. Then

e may be represented by an idempotent $n \times n$ matrix whose entries are in $\mathbb{Q}_d$. Then there is an $\mathcal{S}^d$-isomorphism $f: W \to W$ such that $f^{-1} e f$ is diagonal with entries 0 or 1. Then the retract representing $f^{-1} e f$ is isomorphic to a wedge of spheres, hence the same is true of the retract representing e.

If more than one dimension is represented in W, then write $W = W_1 \vee W_2$ where all the top and only the top dimensional spheres appear in W_2. So $\{W_1, W_2\} = 0$.

Write $e = \begin{pmatrix} e_{11} & e_{12} \\ 0 & e_{22} \end{pmatrix}$, $e_{ij}: W_j \to W_i$. Now

$$\begin{pmatrix} e_{11} & e_{12} \\ 0 & e_{22} \end{pmatrix} = e = e^2 = \begin{pmatrix} e_{11}^2 & e_{11}e_{12} + e_{12}e_{22} \\ 0 & e_{22}^2 \end{pmatrix}$$

so e_{11} and e_{22} are idempotent. Thus the retracts of W_1 and W_2 represented by them are wedges of spheres (e_{22} be the first part, e_{11} by induction). But letting

$f = \begin{pmatrix} 1 & -e_{12} \\ 0 & 1 \end{pmatrix}$ an automorphism we see that

$$\mathrm{Im}(e) \cong \mathrm{Im}(f^{-1} e f) = \mathrm{Im} \begin{pmatrix} e_{11} & 0 \\ 0 & e_{22} \end{pmatrix} = \mathrm{Im}(e_{11}) \oplus \mathrm{Im}(e_{22})$$

a wedge of spheres.

Given a finite CW complex X and a dimension n there is a maximal bouquet of n-spheres B_n which is a retract of X. (Clearly

B_n can have no more n-spheres than B_X since $H_n(B_n)$ is a summand of $H_n(X))$. Let $B_n \xrightarrow{f_n} X \xrightarrow{g_n} B_n = 1$ be the map. Then $B^X = \bigvee_n B_n$ is called the total spherical retract of X. $B^X \xrightarrow{(f_1,\ldots)} X \xrightarrow{\begin{pmatrix} g_1 \\ \vdots \end{pmatrix}} B^X = 1 + \alpha$ where α is nilpotent, hence $1 + \alpha$ is an equivalence. Hence B^X is a retract of X. In general $B^X \neq B_X$.

<u>Lemma 4.31</u>: If S^r retracts from X and $X \equiv Y$ then S^r retracts from Y.

<u>Proof</u>: By suspension we may as well assume $r = 0$. Since $X \equiv Y$, $X \vee B \simeq Y \vee B$, where $B = B_X = B_Y$. Write $B = B_o \vee B'$ where B_o is a wedge of 0-spheres and B' has only spheres of other dimensions. Then $B_o \vee S^o$ retracts from

$$\begin{array}{c} X \vee B_o \vee B' \\ \wr\wr \\ Y \vee B_o \vee B' \end{array}$$

Now look at the identity as the composite

$$(B_o \vee S^o) \xrightarrow{\begin{pmatrix} a \\ a' \end{pmatrix}} (Y \vee B_o) \vee B' \xrightarrow{(b,b')} (B_o \vee S^o)$$

Thus $1 = ba + b'a'$.

But any map $S^o \to S^n \to S^o$ for any $n \neq 0$ is 0. So $b'a' = 0$ so $ba = 1$.

Thus $B_o \vee S^o$ retracts from $Y \vee B_o$. In what follows Y^n is the n-skeleton of Y. Now

$$(S^o \vee B_o \to Y \vee B_o) = (S^o \vee B_o \to Y^o \vee B_o \subset Y \vee B_o)$$

and

$$(Y^{o} \vee B_{o} \subset Y \vee B_{o} \to S^{o} \vee B_{o}) = (Y^{o} \vee B_{o} \to Y^{o}/Y^{-1} \vee B_{o} \to S^{o} \vee B_{o})$$

since $[S^{o} \vee B_{o}, Y/Y^{o}] = 0$ and $[Y^{-1}, S^{o} \vee B^{o}] = 0$. I.e.,

$$\begin{array}{ccccccc} & & Y^{o} \vee B_{o} & \longrightarrow & Y^{o}/Y^{-1} \vee B_{o} & = & \bigvee_{J_{o}} S^{o} \vee B_{o} \\ & \nearrow & \downarrow & & \downarrow & & \\ S^{o} \vee B_{o} & \longrightarrow & Y \vee B_{o} & \longrightarrow & S^{o} \vee B_{o} & = & 1_{S^{o} \vee B_{o}} \end{array}$$

is commutative. Thus at least one of the S^{o} in $S^{o} \vee B_{o}$ must retract from $\bigvee_{J_{o}} S^{o} \vee B_{o}$. So pick automorphisms of $S^{o} \vee B_{o}$ so that S^{o} retracts from $\bigvee_{J_{o}} S^{o} \vee B_{o}$ under these maps. But then S^{o} retracts from Y .

Then picking off one sphere at a time we get

<u>Theorem 4.32</u>: If $X \equiv Y$ then they have the same total spherical retract; i.e. $B^{X} = B^{Y}$.

<u>Proposition 4.33</u>: If $X \vee B \equiv Y$, B a bouquet such that B_{X} is a retract of B , then $X \vee B \simeq Y$.

<u>Proof</u>: B is a retract of $X \vee B$ hence by Theorem 4.29, B is a retract of Y so $Y \simeq B \vee Y'$. Then $X \equiv Y'$. Let $B \simeq B_{X} \vee B'$. Since $X \equiv Y'$ we have $X \vee B_{X} \simeq Y' \vee B_{X}$ so
$X \vee B \simeq X \vee B_{X} \vee B' \simeq Y' \vee B_{X} \vee B' \simeq Y' \vee B \simeq Y$.

<u>Lemma 4.34</u>: If $f,g : X \to Z$ are of orders m,n respectively, where $(m,n) = d$ then $\operatorname{im} f \oplus \operatorname{im} g$ and $\operatorname{im}(f+g)$ are mod d equivalent. In particular, if $(m,n) = 1$ then $\operatorname{im}(f+g) \cong \operatorname{im} f \oplus \operatorname{im} g$.

<u>Proof</u>: Let $sm + tn = d$. Then

$$\begin{array}{ccc} (X & \xrightarrow{f+g} & Z) \\ {\scriptstyle \begin{pmatrix} d \\ d \end{pmatrix}}\downarrow & & \downarrow{\scriptstyle \begin{pmatrix} tn \\ sm \end{pmatrix}} \\ (X\vee X & \xrightarrow{f\vee g} & Z\vee Z) \end{array} \qquad \text{and} \qquad \begin{array}{ccc} (X\vee X & \xrightarrow{f\vee g} & Z\vee Z) \\ {\scriptstyle (tn,sm)}\downarrow & & \downarrow{\scriptstyle (d,d)} \\ (X & \xrightarrow{f+g} & Z) \end{array}$$

commute and the composites are

$$\begin{array}{ccc} (X & \xrightarrow{f+g} & X) \\ {\scriptstyle d^2}\downarrow & & \downarrow{\scriptstyle d^2} \\ (X & \xrightarrow{f+g} & Z) \end{array} \qquad \text{and} \qquad \begin{array}{ccc} (X\vee X & \xrightarrow{f\vee g} & Z\vee Z) \\ {\scriptstyle d\begin{pmatrix} tn & sm \\ tn & sm \end{pmatrix}}\downarrow & & \downarrow{\scriptstyle d\begin{pmatrix} tn & tn \\ sm & sm \end{pmatrix}} \\ (X\vee X & \xrightarrow{f\vee g} & Z\vee Z) \end{array}$$

But $(f\vee g)\begin{pmatrix} tn & sm \\ tn & sm \end{pmatrix} = \begin{pmatrix} f & 0 \\ 0 & g \end{pmatrix}\begin{pmatrix} tn & sm \\ tn & sm \end{pmatrix} = \begin{pmatrix} df & 0 \\ 0 & dg \end{pmatrix} = df \vee dg$,

$= d(f\vee g)$ so the second composite is d^2 times the identity and so is the first. Since $(X \xrightarrow{f+g} Z) = \operatorname{im}(f+g)$ and $(X \vee X \xrightarrow{f\vee g} Z \vee Z) = \operatorname{im} f \oplus \operatorname{im} g$, the Lemma is proved.

4.4 Primary Spaces

Definition: Let $T(X) \geq 0$ generate the ideal in Z $\{m \mid$ there are maps $X \to B \to X = m1_X$, B a bouquet of spheres$\}$. Then let $\mathbb{P}_X = \{\text{primes } p \mid p \mid T(X)\}$. We say X is p-primary if and only if $\mathbb{P}_X = \{p\}$. $\mathbb{P}_X = \emptyset$ if and only if $X = B_X$. (Since $\mathbb{P}_X = \emptyset$ implies X is a retract of a wedge of spheres, hence is itself a wedge of spheres.)

Example: Let $\theta \in \pi_{r-1}(\underline{S})$ be of order p^n . Then $C_\theta = S^o \cup_\theta e^r$ is p-primary.

Proof: We have $S^o \xrightarrow{i} C_\theta \xrightarrow{\sigma} S^r$. From the cofibration properties, there are maps v and u making the following commute

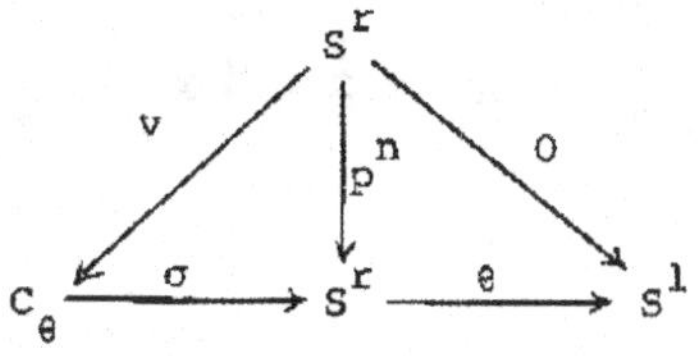

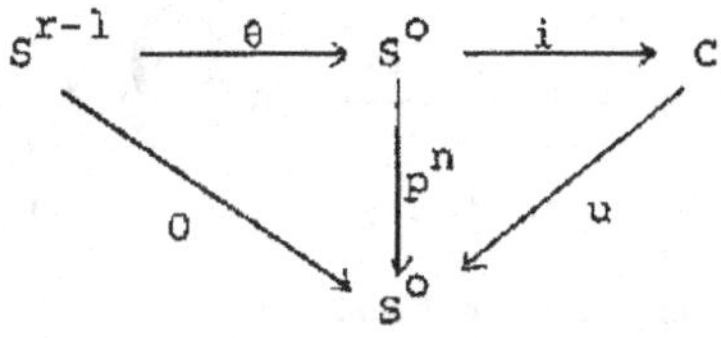

Thus we have $C_\theta \xrightarrow{\binom{u}{\sigma}} S^o \vee S^r \xrightarrow{(iv)} C_\theta$; let $h = iu + v\sigma$. Then $hi = iui = ip^n = p^n i$ and $\sigma h = v\sigma = p^n \sigma = \sigma p^n$. Thus $(h - p^n)i = 0$, $\sigma(h - p^n) = 0$. Thus $h - p^n = k\sigma$. Let $v' = v - k$. Then $h' = (iv')\binom{u}{\sigma} = iu + v'\sigma = iu + v\sigma - k\sigma = h - k\sigma = p^n$.

Let us observe that in fact $T(X) > 0$: if we let $\mathcal{S}^o$ be as before, i.e. it has the same object as $\mathcal{S}$ with $\mathcal{S}^o(X,Y) = \mathcal{S}(X,Y) \otimes \mathbb{Q}$ then $f: X \to Y$ is an isomorphism in $\mathcal{S}^o$ if and only if there is some map $g: Y \to X$ such that $fg = m1_Y$ and $gf = m1_X$ for some integer

$m \neq 0$. Looking at $H_*(C_f;Q)$ shows us that $H_*(f;Q)$ is an isomorphism if and only if f is an isomorphism in $\mathcal{S}^0$. (The flatness of Q tells us that exact sequences remain exact when tensored with Q .)

Now let $\varphi: B_X \to X$ be an $F\pi_*$-isomorphism. Then, in particular, $H_*(\varphi;Q)$ is an isomorphism so φ is an $\mathcal{S}^0$ isomorphism so there is some $\theta: X \to B_X$ with $\varphi\theta = m1_X$ for some integer $m > 0$. Thus $T(X) > 0$.

We observe easily that $T(X \vee Y)$ = least common multiple $\ell.c.m.(T(X),T(Y))$, whence $\mathbb{P}_{X\vee Y} = \mathbb{P}_X \cup \mathbb{P}_Y$: let $t_1 = T(X)$, $t_2 = T(Y)$, $t = T(X \vee Y)$, $m = \ell.c.m.(t_1,t_2)$, $d = g.c.d.(t_1,t_2)$. $dm = t_1t_2$. We wish to show that $m = t$.

If $X \vee Y \xrightarrow{(f,f')} B \xrightarrow{\binom{g}{g'}} X \vee Y = t1_{X\vee Y}$ then $X \xrightarrow{f} B \xrightarrow{g} X = t1_X$ so $t_1 | t$. Similarly $t_2 | t$ so $m | t$.

On the other hand if

$$X \xrightarrow{f} B_1 \xrightarrow{g} X = t_1 1_X$$

and

$$Y \xrightarrow{f'} B_2 \xrightarrow{g'} Y = t_2 1_X$$

then

$$X \vee Y \xrightarrow{\frac{t_2}{d} f \vee \frac{t_1}{d} f'} B_1 \vee B_2 \xrightarrow{g \vee g'} X \vee Y = m1_{X\vee Y}$$

so $t | m$. Thus $t = m$.

We also observe the following: in $\mathcal{S}^{T(X)}$ we have

$X \xrightarrow{f} B \xrightarrow{\frac{g}{T(X)}} X = 1_X$ where $gf = T(X)1_X$, so X is a retract of B

in $\mathcal{S}^{T(X)}$ hence $X \simeq B'$ in $\mathcal{S}^{T(X)}$ (from Theorem 4.28) where B' is a wedge of spheres. By taking homology groups, it is easy to see that $B' \simeq B_X$. Thus $X \simeq B_X$ in $\mathcal{S}^{T(X)}$. That is, there exist maps $\varphi : X \to B_X$, $\psi : B_X \to X$ such that $\psi\varphi = m1_X$, $\varphi\psi = m1_{B_X}$ where all the primes of m are in $\mathbb{P}_X$. Thus in any category where the primes of $\mathbb{P}_X$ are invertible we have $X \simeq B_X$.

Example: It is interesting to note that we can not necessarily get maps $X \to B_X \to X = T(X)1_X$: Let $X = S^o \cup_2 e^1$. We show in Chapter 5 that 1_X is of order 4 and that the following diagram commutes

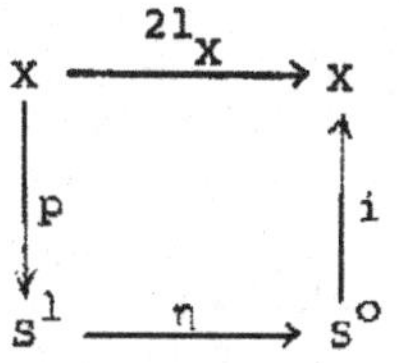

for a map $\eta \in \pi_1(S^o)$. Thus $X \xrightarrow{p} S^1 \xrightarrow{i\eta} X = 21_X$ so $T(X) = 2$. But since X is torsion, $B_X = *$ and thus $X \to B_X \to X = 0 \neq 21_X$.

Assume X is p-primary and $X \xrightarrow{\alpha} B' \xrightarrow{\beta} X = p^m 1_X$ where B' is a bouquet of spheres. Let $\varphi : B \to X$ be an $F\pi_*$-isomorphism. Then, since B' is a bouquet of spheres, β can be factored through φ up to torsion. That is, for some map $a : B' \to B$

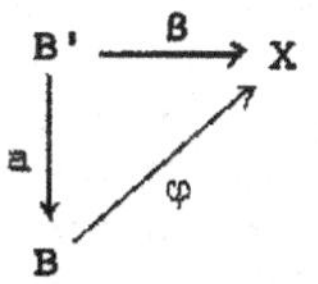

commutes up to torsion; i.e. there is a map $b : B' \to X$ of finite order p^{n-m}, where $p \nmid d$, and $\varphi a - \beta = b$. Then if $\beta' = p^{n-m}\beta$,

$a' = p^{n-m}a$ and $b' = p^{n-m}b$, we have $X \xrightarrow{\alpha} B' \xrightarrow{\beta'} X = p^n 1_X$ and $\varphi a' - \beta' = b'$ where b' is of order d, $p \nmid d$. Since $(p^n, d) = 1$ there are integers s and t such that $sp^n + td = 1$. Then $sp^n b' = b'$. Let $\gamma = s\alpha b' : B' \to B'$. $\beta'\gamma = s\beta'\alpha b' = sp^n b' = b' = \varphi a' - \beta'$. Thus $\beta'(1_{B'} + \gamma) = \varphi a'$. But γ is a torsion element $B' \to B'$, hence if we write it in matrix form in terms of a splitting of B' into spheres in increasing order, the matrix is strictly upper triangular, hence is nilpotent. Thus $1_{B'} + \gamma$ is invertible (its inverse is $g' = 1_{B'} - \gamma + \gamma^2 - \gamma^3 + \ldots$). Thus $\beta' = \beta'(1_{B'} + \gamma)g' = \varphi a'g'$. Thus letting $g = a'g' : B' \to B$ we have

$$X \xrightarrow{\alpha} B' \xrightarrow{\beta'} X = p^n 1_X$$

and

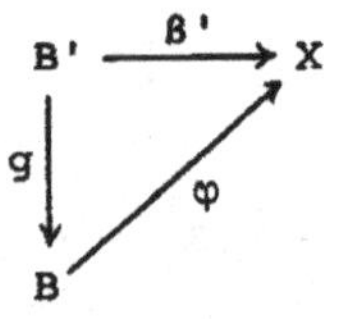

commutative. Thus if $i : X \to C$ is the cokernel of φ we have $p^n i = i\beta'\alpha = i\varphi g\alpha = 0$ since $i\varphi = 0$. Thus since i is onto, C is p-torsion. Let $K = \ker \varphi$. Then we have the exact sequence

$$0 \to C \to C_\varphi \to SK \to 0 .$$

Since φ is an $H_*(\ ;Q)$ isomorphism C_φ is a torsion space. Let $C_\varphi = C_p \vee C'$ where C_p is the p-primary part of C_φ. Then C maps into C_p so C' maps monomorphically to SK. But SK maps monomorphically to SB so there is a monomorphism $C' \to SB$. Since C' is injective, it is a summand of SB - a bouquet of spheres. But the only torsion space that can be a summand of a bouquet of spheres is a trivial space. Thus C' is trivial so $C_\varphi = C_p$ is p-torsion. Thus SK and hence K is p-torsion. Thus

<u>Proposition 4.35</u>: If X is p-primary and $\varphi: B \to X$ an $F\pi_*$-isomorphism, then ker φ and cok φ are p-torsion objects.

Dually we have

<u>Proposition 4.35D</u>: If X is p-primary and $\psi: X \to B$ is an $F\pi^*$-isomorphism then ker ψ and cok ψ are p-torsion.

Thus φ and ψ are isomorphisms in $\mathcal{S}^P$, hence so is $\psi\varphi$. Now look at the restriction of ψ and φ to B_n where B_n is the wedge of all the n-spheres in B. The restriction $\psi_n: X \to B_n$ is an $F\pi^n$-isomorphism and $\varphi_n: B_n \to X$ is an $F\pi_n$-isomorphism. $\psi_n\varphi_n$ is an isomorphism in $\mathcal{S}^P$ hence we can diagonalize its matrix and all diagonal entries are units of $\mathbb{Q}^P$, i.e. are integers of the form p^n, $n \geq 0$. If we assume that there is no map $S^n \to X \to S^n = 1_{S^n}$, then all the non-zero entries of $\psi\varphi$ are of the form p^n, $n \geq 1$, hence $\psi_n\varphi_n$ is divisible by p. Let $\psi_n\varphi_n = pf$.

Now let $S^n \xrightarrow{h} X \xrightarrow{k} S^n$ be any map, then since φ_n and ψ_n are $F\pi_n$ and $F\pi^n$ isomorphisms, respectively, there are maps $\tilde{h}: S^n \to B_n$ and $\tilde{k}: B_n \to S^n$ such that

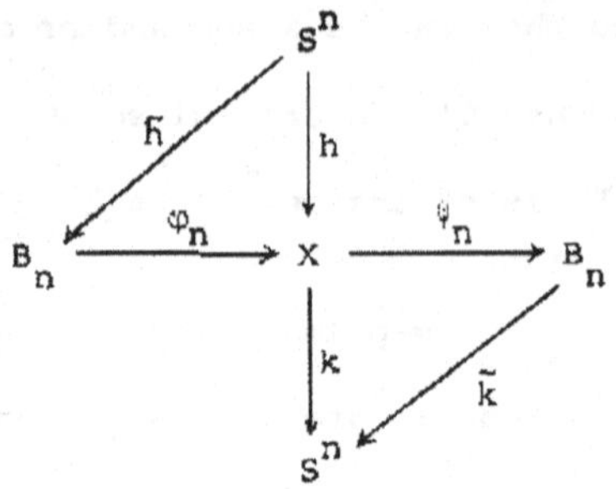

commutes up to torsion. Thus $kh - \tilde{k}\psi_n\varphi_n\tilde{h} = kh - \tilde{k}(pf)\tilde{h}$ is of finite order, i.e. $kh - p(\tilde{k}f\tilde{h}):S^n \to S^n$ is of finite order. But End $S^n = Z$ so $kh = p(\tilde{k}f\tilde{h})$. Thus

Theorem 4.36: If X is p-primary and has no spherical retracts of dimension n then any map $S^n \to X \to S^n$ is divisible by p .

Corollary 4.37: If X and Y are p-primary and have no spherical retracts of degree n , then neither does $X \vee Y$.

Proof: If $S^n \xrightarrow{\binom{f}{g}} X \vee Y \xrightarrow{(h,k)} S^n$ is any map, then by the Theorem, hf and kg are each divisible by p , hence $hf + kg$ is divisible by p .

Lemma 4.38: Assume there are maps $\binom{f}{g}:X \to Y \vee Z$ and $(f',g'):Y \vee Z \to X$ whose composite is $k1_X$, k an integer such that $p \nmid kT(Z)$ where X and Y are p-primary. Then if $B^X = *$, X congruence retracts from Y (i.e. $X \vee A$ retracts from $Y \vee A$ for some space A).

Proof: Let

$$Z \xrightarrow{a} B_1 \xrightarrow{b} Z = T(Z)1_Z$$

and

$$X \xrightarrow{a'} B_2 \xrightarrow{b'} X = T(X)1_X \,.$$

Then for any integers m and s we can form

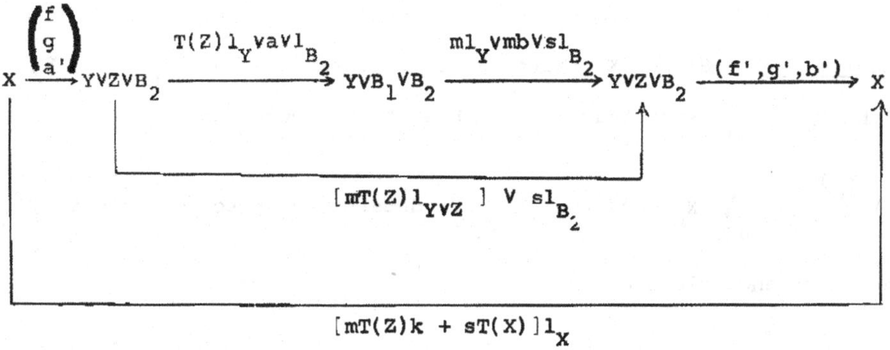

In particular, choose m and s so that $mT(Z)k + sT(X) = 1$. Thus X retracts from $Y \vee B$ where B is a bouquet of spheres. Thus $X \vee W \simeq Y \vee B$. Writing $W = W' \vee B^W$ we have $X \vee W' \vee B^W \simeq Y \vee B$. $\mathbb{P}_{W'} \subseteq \{p\}$ so from Corollary 4.37, $X \vee W'$ has no spherical retracts. Using Theorem 4.32 and pulling off one sphere at a time, it is clear that the total spherical retract of the left hand side is B^W and that of the right hand side is $B^Y \vee B$. Thus $B^W = B^Y \vee B$ so $X \vee W' \vee B^Y \vee B \simeq Y \vee B$. Thus $X \vee B$ is a retract of $Y \vee B$.

Taking $k = 1$ in the above yields

<u>Corollary 4.39</u>: Assume X and Y are p-primary and Z is such that $p \notin \mathbb{P}_Z$. If X has trivial spherical retract and X is a retract of $Y \vee Z$ then X is a congruence retract of Y.

Finally we are ready to state explicitly why G is free, i.e.

what the unique generators X_α are such that given a finite CW complex X, $[X] = \Sigma n_\alpha [X_\alpha]$ in G - i.e. allowing repetition we get $X \vee X_{\alpha_1} \vee X_{\alpha_2} \vee \ldots \vee X_{\alpha_n} \equiv X_{\alpha_{n+1}} \vee \ldots \vee X_{\alpha_{n+m}}$, uniquely. Actually the result is a bit stronger. Most of what we need is the following:

Theorem 4.40: Given X there exist p-primary spaces X_p, $p \in \mathbb{P}_X$ with $B_{X_p} = B_X$ such that if $\mathbb{P}_X$ has n elements, then

$X \vee \bigvee_1^{n-1} B_X \simeq \bigvee_{p \in \mathbb{P}_X} X_p$ and the X_p are determined up to congruence by the congruence class of X.

Proof: Let $\varphi: B_X \to X$ be an $F\pi_*$-isomorphism. Let Y be such that $Y \xrightarrow{f} B_X \xrightarrow{\varphi} X$ is a cofibration. Then Y is torsion. Then End Y is a finite ring, hence is the direct sum of its p-primary components - a component for each element of $\mathbb{P}_Y$. Each component is, of course, idempotent generated so $Y = \bigvee_{p \in \mathbb{P}_Y} Y_p$. Let $f_p = f|Y_p : Y_p \to B_X$ and let X_p be the mapping cone. Using Corollary 4.24 we have

$$X \vee \bigvee_1^{n-1} B_X \simeq \bigvee_{p \in \mathbb{P}_Y} X_p .$$

Clearly $B_X = B_{X_p}$.

First let us observe that each X_p is p-primary: let $\exp Y_p = p^n$, $f_p : Y_p \to B_X$. Clearly $B_{X_p} = B_X$. $i_{f_p} : B_X \to X_p = C_{f_p}$ is one map and since $\sigma_{f_p}(p^n 1_{B_X}) = p^n \sigma_{f_p} = 0$ there is a lifting $g: X_p \to B_X$ with $i_{f_p} \circ g = p^n 1_{B_X}$. Thus X_p is either p-primary or

$X_p \simeq B_X$. Since $Y_p \neq *$, X_p is p-primary.

Then $\mathbb{P}_X = \mathbb{P}(X \vee \bigvee_1^{n-1} B_X) = \mathbb{P}(\bigvee_{p\in\mathbb{P}(Y)} X_p) = \bigcup_{p\in\mathbb{P}(Y)} \mathbb{P}(X_p) = \mathbb{P}_Y$.

Let $\mathbb{P} = \mathbb{P}_X$. For the uniqueness of the X_p we assume $\bigvee_{p\in\mathbb{P}} X_p \simeq \bigvee_{p\in\mathbb{P}} X'_p$ and $B_{X_p} = B_X = B_{X'_p}$. Fixing p, write $X_p = X_1 \vee B_1$ and $X'_p = X_2 \vee B_2$ where X_1 and X_2 have trivial spherical retracts. Then $X_1 \vee (B_1 \vee \bigvee_{q\neq p} X_q) \simeq X_2 \vee (B_2 \vee \bigvee_{q\neq p} X'_q)$. Then by Corollary 4.39, X_1 is a congruence retract of X_2 and X_2 is a congruence retract of X_1, hence $X_1 \equiv X_2$. But then $B_{X_1} = B_{X_2}$ and since $B_{X_1} \vee B_1 = B_{X_p} = B_{X'_p} = B_{X_2} \vee B_2$, $B_1 = B_2$ so $X_p \equiv X'_p$. Thus we have the uniqueness completing Theorem 4.40.

4.5 The Category $\mathcal{S}_p$

We shall localize $\mathcal{S}$ in a different way. Let p be a fixed prime and let d be the set of all primes other than p. Then let $\mathcal{S}_p$ be $\mathcal{S}^d$. Thus $\mathcal{S}_p$ is the category in which all primes other than p become units: $\mathbb{Q}_p = \mathbb{Q}^d$, the set of rational numbers $\frac{m}{n}$ with $(n,p) = 1$; and $\mathcal{S}_p(X,Y) = \mathcal{S}(X,Y) \otimes \mathbb{Q}_p$. The main properties of $\mathbb{Q}_p$ are

1) $\mathbb{Q}_p$ is local: $p\mathbb{Q}_p$ is the ideal of non-units;
2) $\mathbb{Q}_p$ is Z flat; i.e. tensoring with $\mathbb{Q}_p$ preserves exactness;
3) $Z_m \otimes \mathbb{Q}_p \cong Z_{p^n}$ where p^n is the highest power of p dividing m.

Observe that if $p \notin \mathbb{P}_X$, then $X \simeq B_X$ in $\mathcal{S}_p$. Also note that $\mathrm{End}(S^n) \simeq \mathbb{Q}_p$ is local. Thus $S^n \vee X \simeq S^n \vee Y$ in $\mathcal{S}_p$ implies $X \simeq Y$ in $\mathcal{S}_p$. By induction, if B is a wedge of spheres then $B \vee X \simeq B \vee Y$ in $\mathcal{S}_p$ implies $X \simeq Y$ in $\mathcal{S}_p$. In particular, then, if $X \equiv Y$ then $X \vee B_X \simeq Y \vee B_X$ in $\mathcal{S}$, hence in $\mathcal{S}_p$, so $X \simeq Y$ in $\mathcal{S}_p$. This proves half of

<u>Theorem 4.41</u>: If $X \equiv Y$ then $X \simeq Y$ in $\mathcal{S}_p$ for every p; conversely if $X \simeq Y$ in $\mathcal{S}_p$ for all $p \in \mathbb{P}_{X \vee Y}$ and $B_X = B_Y$, then $X \equiv Y$.

<u>Proof</u>: First we point out that if $X \simeq Y$ in $\mathcal{S}_p$ for some p then for homology reasons $B_X = B_Y$ so the last condition is in case $\mathbb{P}_{X \vee Y} = \emptyset$. (E.g. $S^n \simeq S^n \vee S^n$ in $\mathcal{S}_p$ for every $p \in \mathbb{P}_{S^n \vee S^n \vee S^n} = \emptyset$.)

Since $X \simeq B_X = B_Y \simeq Y$ for all $p \notin \mathbb{P}_{X \vee Y}$, we may assume given that $X \simeq Y$ in $\mathcal{S}_p$ for all p. Thus $X \simeq_{m_p} Y$ for some integer m_p

with $(p,m_p) = 1$. Fix p_1 . Let $p_2,\ldots,p_n$ be the primes dividing m_{p_1} . Then $m_{p_1}, m_{p_2},\ldots,m_{p_n}$ have greatest common divisor 1 . Then this theorem follows from this result:

<u>Proposition 4.42</u>: Let $m_1,\ldots,m_n$ be integers with greatest common divisor 1 . Assume $X \simeq_{m_i} Y$ for $i = 1,\ldots,n$. Then $X \equiv Y$.

<u>Proof:</u> Let $f_i:X \to Y$, $g_i:Y \to X$ be such that $g_i f_i = m_i 1_X$ and $f_i g_i = m_i 1_Y$. (Actually we only know tnat such maps exist with composite m_i^N for some N , but we may as well replace our original m_i with m_i^N in this case as $\simeq_{m_i}$ and $\simeq_{m_i^N}$ mean the same.)

Let $S^{-1}C_{f_i} \xrightarrow{\sigma_i} X \xrightarrow{f_i} Y \xrightarrow{j_i} C_{f_i}$ be the usual maps. $m_i j_i = j_i m_i = j_i f_i g_i = 0$ and $m_i \sigma_i = g_i f_i \sigma_i = 0$. Let $Y \overset{\alpha}{\to} B \overset{\beta}{\to} Y = T(Y)1_Y$. Then $j_i(m_i\beta) = 0$ so there is a map $\tilde{\beta}:B \to X$ such that $f_i\tilde{\beta} = m_i\beta$. Then $f_i[\tilde{\beta}(\alpha f_i) - m_i T(Y)1_X] = m_i T(Y) f_i - m_i T(Y) f_i = 0$. So $\beta\alpha f_i - m_i T(Y)1_X = \sigma_i k$ for some $k:X \to S^{-1}C_{f_i}$. Thus $(m_i\tilde{\beta})(\alpha f_i) - m_i^2 T(Y)1_X = m_i\sigma_i k = 0$ so

$$X \xrightarrow{\alpha f_i} B \xrightarrow{m_i\tilde{\beta}} X = m_i^2 T(Y)1_X .$$

Thus $T(X)|m_i^2 T(Y)$ for all i . Thus $T(X)| \gcd_{i=1,\ldots,n} (m_i^2 T(Y)) = \gcd_{i=1,\ldots,n} (m_i^2)T(Y) = T(Y)$. By symmetry $T(Y)|T(X)$. Thus $T(X) = T(Y)$. Let $\mathbb{P}_X = \mathbb{P}_Y = \{p_1,\ldots,p_t\}$. Then let $\bar{B} = \bigvee_1^{t-1} B_X$ $X \vee \bar{B} = \bigvee_1^t X_{p_i}$, $Y \vee \bar{B} = \bigvee_1^t Y_{p_i}$. Then by mapping $\bar{B} \to \bar{B}$ by the identity (respectively by $m_i 1_{\bar{B}}$) we can extend the f_i (resp. the g_i) to maps between $\bigvee X_{p_i}$ and $\bigvee Y_{p_i}$. For each p choose m_i so that

$p \not| m_i$. Write $X_p \equiv X'_p \vee B'$ and $Y_p \equiv Y'_p \vee B''$ where B' and B'' are the total spherical retracts of X_p and Y_p. Clearly $B_{X'_p} \vee B' = B_{X_p} = B_{Y_p} = B_{Y'_p} \vee B''$ so that if we prove $X'_p \equiv Y'_p$ then it will follow that $B' = B''$ so that $X_p \equiv Y_p$. Now X'_p is a retract of $\bigvee X_q = \bigvee Y_q = Y'_p \vee B'' \vee \bigvee_{q \neq p} Y_q$, so by Cor. 4.39, X'_p is a congruence retract of Y'_p. Similarly Y'_p is a congruence retract of X'_p. Hence $X'_p \equiv Y'_p$ and thus $X_p \equiv Y_p$. So $X \vee \bar{B} = \bigvee X_p \equiv \bigvee Y_p = Y \vee \bar{B}$ so $X \equiv Y$ and the Proposition is proved.

Apply Theorem 4.40 to a space X yielding $X \vee \bigvee_1^{n-1} B_X \simeq \bigvee_{p \in \mathbb{P}_X} X_p$ where $\mathbb{P}_X$ has n elements. Then in $\mathcal{S}_p$ we have $X_q \simeq B_{X_q} = B_X$ for all $q \in \mathbb{P}_X - \{p\}$. Thus in $\mathcal{S}_p$, $X \vee \bigvee_1^{n-1} B_X \simeq X_p \vee \bigvee_1^{n-1} B_X$ so $X \simeq X_p$ in $\mathcal{S}_p$, since we have already observed that in $\mathcal{S}_p$, we can cancel bouquets of spheres.

<u>Corollary 4.43</u>: If X has no spherical retracts in $\mathcal{S}$ and X is p-primary, then X has no spherical retracts in $\mathcal{S}_p$.

<u>Proof</u>: Assume $X \simeq S^n \vee Y$ in $\mathcal{S}_p$. Applying tne above argument, we find a unique (up to congruence) space Y_p with $Y \simeq Y_p$ in $\mathcal{S}_p$. Then $X \simeq S^n \vee Y_p$ in $\mathcal{S}_p$. Applying Theorem 4.41, we get that $X \equiv S^n \vee Y$. But Theorem 4.32 says that total spherical retracts depend only on the congruence class, so S^n retracts from X.

4.6 $K_o(\mathcal{S})$ is free

We shall now prove Theorem 4.15 that $G = K_o(\mathcal{S})$, the Grothendieck group, is free, and we can now say what the generators are.

Theorem 4.44: G is free with generators spheres of all dimensions and for all p indecomposable p-primary spaces.

The remainder of the section will be devoted to the proof of this. Let G_F be the subgroup of G generated by all spheres. Let G_p be the subgroup of G generated by indecomposable p-primary spaces. An element of G_F can be written as $[B] - [B]$ where B and B' are bouquets of spheres. An element of G_p can be written as $[X] - [X']$ where X and X' are p-primary spaces with no spherical retract (this uses Corollary 4.37).

First we observe that G_F is clearly free. We will soon prove that G_p is free on the indecomposable p-primary spaces. Then the main theorem will be proved since

Lemma 4.45: $G = G_F \oplus \bigoplus_p G_p$.

Proof: First we observe that for any space X , Theorem 4.40 can be interpreted as saying that $[X] = \Sigma[X_p] - (n-1)[B_X]$. $[B_X] \in G_F$ and $[X_p] \in G_F + G_p$, hence $G = G_F + \sum_p G_p$; so we must prove that this sum is direct.

Assume $0 = [B] - [B'] + \sum_p [X_p] - [X'_p]$ where B and B' are bouquets of spheres and X_p and X'_p p-primary spaces with trivial spherical retract. Then $Z = B \vee \bigvee_p X_p \equiv B' \vee \bigvee_p X'_p = Z'$. Now $\{p | X_p \neq *\} = \mathbb{P}(B \vee \bigvee_p X_p) = \mathbb{P}(B' \vee \bigvee_p X'_p) = \{p | X'_p \neq *\}$. So we may take the sum over the finite set $\mathbb{P}$ of primes p for which X_p and X'_p are non-trivial; assume $\mathbb{P}$ has n elements. By adding some non-trivial spaces to each side, we may assume that $n \geq 2$. Let $Y_p =$ $X_p \vee B \vee \bigvee_{q \neq p} B_{X_q}$, $Y'_p = X'_p \vee B' \vee \bigvee_{q \neq p} B_{X'_q}$. Then

$B_{Y_p} = B_{X_p} \vee B \vee \bigvee_{q \neq p} B_{X_q} = B_Z = B_{Z'} = B_{Y'_p}$. The decomposition for Z and Z' yields $\bigvee_p Y_p \simeq Z \vee \bigvee_1^{n-1} B_Z \simeq Z' \vee \bigvee_1^{n-1} B_{Z'} \simeq \bigvee_p Y'_p$. The middle $\simeq$ comes from the assumption that $n \geq 2$ and Proposition 4.33. Then by the uniqueness part of the decomposition theorem, $Y_p \equiv Y'_p$. That is, $X_p \vee B \vee \bigvee_{q \neq p} B_{X_q} \equiv X'_p \vee B' \bigvee_{q \neq p} B_{X'_q}$. By the dependence of spherical retracts on the congruence class, we get $X_p \equiv X'_p$ since X_p and X'_p have trivial spherical retract. Thus $B_{X_p} = B_{X'_p}$ for all p , hence $B = B'$. Thus $[B] - [B'] = 0$ and $[X_p] - [X'_p] = 0$. Thus the sum is direct. Thus the lemma is proved.

It will now suffice to show that G_p is free on generation $[W]$ where W is an indecomposable p-primary space with trivial spherical retract. Assume, for the moment, that for such a W , $\mathrm{End}_{\mathfrak{g}_p}(W)$ is local. Then given $\bigvee W_i \equiv \bigvee W'_j$ where all the W_i, W'_j are p-primary indecomposable spaces, we get $\bigvee W_i \simeq \bigvee W'_j$ in $\mathfrak{g}_p$ and since $\mathrm{End}_{\mathfrak{g}_p}(W)$ is local for $W = W_i$ or W'_j , we get that each W_i is a retract of (and hence equal to) some W'_{j_i} in $\mathfrak{g}_p$. But since they are all p-primary this says that each $W_i \equiv W_{j_i}$ and the decomposition is unique up to congruence class.

Thus Theorem 4.44 will be proved when we show that $\mathrm{End}_{\mathfrak{g}_p}(W)$ is local for W p-primary and indecomposable.

<u>Lemma 4.46</u>: If $f:X \to Y$ and $H_*(f;Z_p)$ is an isomorphism then f is an isomorphism in $\mathcal{S}_p$.

<u>Proof</u>: $H_*(f;Z_p)$ an isomorphism implies that $H_*(C_f;Z_p) = 0$ so C_f has exponent prime to p so $C_f \sim *$ in $\mathcal{S}_p$ so f is an $\mathcal{S}_p$ isomorphism. (This follows from the flatness of $\mathbb{Q}_p$.)

Let $\varphi:B \to X$ be an $F\pi_*$-isomorphism. Let $B = B_X$ with X p-primary. Then $\operatorname{cok} \varphi$ has exponent p^k. Let $n \geq 1$ be large enough so that $p^n \mathcal{S}_p(B,X)$ has no torsion. Then for $\varphi_*:\mathcal{S}_p(B,B) \to \mathcal{S}_p(B,X)$ we have $p^n\varphi_*$ an epimorphism. I.e. $\varphi_*:p^n\mathcal{S}_p(B,B) \to p^n\mathcal{S}_p(B,X)$ is onto. Then if $f \in \mathcal{S}_p(X,X)$, $p^n\varphi^*(f) \in p^n\mathcal{S}_p(B,X)$ so there exists $f' \in \mathcal{S}_p(B,B)$ with $\varphi_*(p^n f') = p^n\varphi^*(f)$. I.e.,

$$\begin{array}{ccc} B & \xrightarrow{p^n\varphi} & X \\ {\scriptstyle f'}\downarrow & & \downarrow{\scriptstyle f} \\ B & \xrightarrow{p^n\varphi} & X \end{array}$$

commutes.

Let $F = \operatorname{cok} p^n\varphi$. Then there is a uniquely defined $f'' \in \operatorname{End}_{\mathcal{S}_p}(F)$ such that

$$\begin{array}{ccccc} X & \longrightarrow & F & \longrightarrow & 0 \\ {\scriptstyle f}\downarrow & & \downarrow{\scriptstyle f''} & & \\ X & \longrightarrow & F & \longrightarrow & 0 \end{array}$$

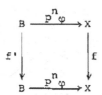

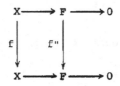

commutes. Hence there is a map $\rho:\mathrm{End}_{\mathcal{G}_p}(X) \to \mathrm{End}_{\mathcal{G}_p}(F)$. Since X is projective this is onto. By its definition, it is a ring homomorphism so it takes units to units.

Claim: ρ(non-unit) is a non-unit.

Proof: Assume $\rho(f)$ is a unit. Let the following diagram commute

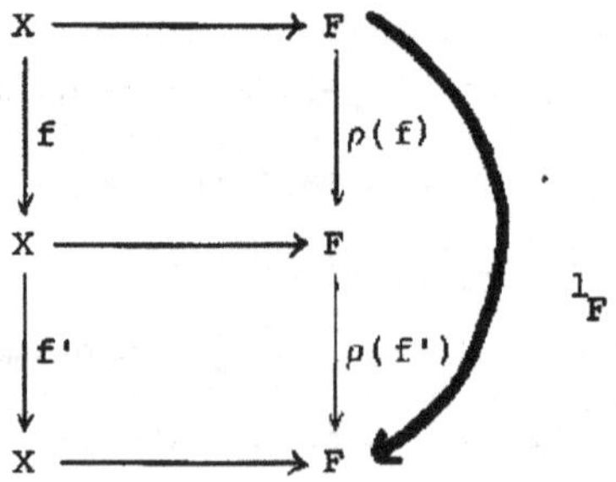

Then since $\rho(f'f) = 1_F$ we have

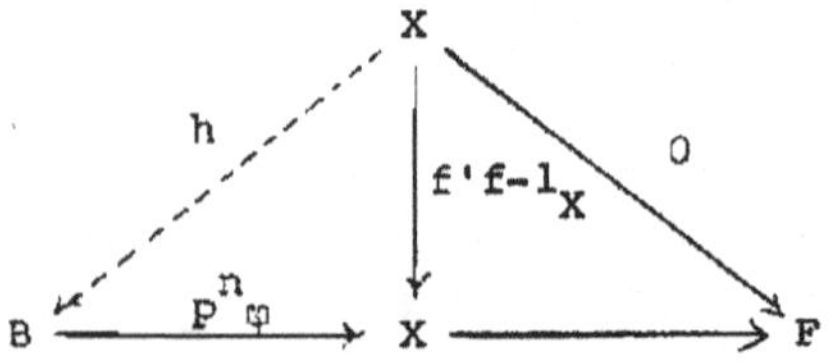

commutative for some h . Thus $f'f - 1_X = p^n\varphi h$ so $f'f = 1_X + p^n\varphi h$. But then $H_*(f'f;Z_p) = H_*(1_X;Z_p) + p^n H_*(\varphi h;Z_p) = 1_{H_*(X;Z_p)}$. Using Lemma 4.46, $f'f$ is an $\mathcal{G}_p$ isomorphism so f is a unit. We now use this to prove

<u>Theorem 4.47</u>: $\mathrm{End}_{\mathcal{G}_p}(X)$ is local if X is p-primary and indecomposable.

<u>Proof</u>: We have just proved that ρ preserves units and non-units, so it will suffice to show that $End_{\mathcal{S}_p}(F)$ is local. Since F is torsion it will suffice to show that $End_{\mathcal{S}_p}(F)$ has no idempotents. We know that $End_{\mathcal{S}_p}(X)$ has no idempotents.

Out aim now is to show that $End_{\mathcal{S}_p}(F)$ has no idempotents. Our method will be as follows: let C be the cofibre of $p^n\varphi$. Then we have the exact sequences

(1) $$0 \to I \to X \to F \to 0$$

(2) $$0 \to F \to C \to K \to 0$$

(3) $$0 \to K \to SB \to SI \to 0$$

(S1) $$0 \to SI \to SX \to SF \to 0$$

where $I = \operatorname{im} p^n\varphi$, $F = \operatorname{cok} p^n\varphi$, $K = \ker p^n\varphi$. We shall show that an idempotent on F leads to one on C and in turn to one on K , then to one on SB and to one on SI and finally to one on SX . Assuming that X is indecomposable, then SX is also, so we get a contradiction.

We also make use of the fact that X is indecomposable to observe that F is reduced - i.e., F has no injective (hence projective) summands: a projective summand of F would, by (1), be a summand of X . Similarly the fact that C is p-torsion implies by (2) that F and K are, so that any injective summand of K is, by (3) a summand of SB . Since a bouquet of spheres can have no torsion summand, K is also reduced. Similarly I is reduced.

Lemma 4.48: Let $0 \to A \overset{p}{\to} B \overset{q}{\to} C \to 0$ be an exact sequence in an a abelian category where summands of injectives are injective with A and C reduced and B injective. Then given idempotents $e:A \to A$ and $f:B \to B$ with $fp = pe$, there is an idempotent $g:C \to C$ with $gq = qf$. If C is torsion then for any idempotent $e:A \to A$ there is an idempotent $f:B \to B$ with $fp = pe$ (and hence an idempotent $g:C \to C$ with $qg = fg$).

We shall prove this lemma soon. But first observe that we can use it to extend an idempotent on F by (1) to one on K. Then, assuming the idempotent on K can be extended to one on SB (this will take much work) we then get an idempotent on SI by (3) and then by (S1) we get one on SX and the needed contradiction.

Thus it remains only to prove Lemma 4.48 and to prove that an idempotent on K can be extended to one on SB.

Proof of Lemma 4.48:

For the first part, we observe that a map $g:C \to C$ is defined such that $gq = qf$. Then $g^2q = qf^2 = qf = gq$. Since q is an epimorphism $g^2 = g$. If $g = 0$, then $fq = 0$, so, by the injectivity of B, there is a map $k:B \to A$ with $pk = f$. But then $pkp = fp = pe$. Since p is a monomorphism $kp = e$. Thus $\operatorname{im} k = \operatorname{im} e$ and $\operatorname{im} e \xrightarrow{p|\operatorname{im} e} B \xrightarrow{k} \operatorname{im} e$ is the identity since $e|\operatorname{im} e$ is the identity. Then $\operatorname{im} e$ is injective which contradicts the fact that A is reduced. If on the other hand $g = 1$ then $g - 1 = 0$ so $\operatorname{im}(e-1)$ is injective yielding the same contradiction. Thus g is an idempotent.

For the second part of the lemma, we are given just $e:A \to A$. Since B is injective, the exact diagram

$$\begin{array}{ccc} 0 \to & A & \xrightarrow{p} B \\ & {\scriptstyle pe}\downarrow & \\ & B & \end{array}$$

can be extended to

$$\begin{array}{cccc} 0 \to & A & \xrightarrow{p} & B \\ & {\scriptstyle pe}\downarrow & \swarrow {\scriptstyle x} & \\ & B & & \end{array}$$

Then the commutativity of

$$\begin{array}{ccccccccc} 0 & \to & A & \to & B & \to & C & \to & 0 \\ & & \downarrow{\scriptstyle e} & & \downarrow{\scriptstyle x} & & & & \\ 0 & \to & A & \to & B & \to & C & \to & 0 \end{array}$$

yields a map $y:C \to C$. Since End C is finite, for some integer n, $y^{2n} = y^n$ (as in the proof of Proposition 4.17). Let $g = y^n$. Then $g^2 = g$. We have the commutative diagram

$$\begin{array}{ccccccccc} 0 & \to & A & \xrightarrow{p} & B & \xrightarrow{q} & C & \to & 0 \\ & & \downarrow{\scriptstyle e} & & \downarrow{\scriptstyle w} & & \downarrow{\scriptstyle g} & & \\ 0 & \to & A & \xrightarrow{p} & B & \xrightarrow{q} & C & \to & 0 \end{array}$$

where $w = x^n$, e is an idempotent and $g^2 = g$.

Now $(w^2 - w)p = p(e^2 - e) = 0$ so there exist $h:C \to B$ such that $w^2 - w = hq$. Similarly $q(w^2 - w) = (g^2 - g)q = 0$. Thus $w^2(w-1)^2 = (w^2 - w)^2 = hq(w^2 - w) = 0$. Furthermore for some integer $n \geq 2$, $nC = 0$ since C is torsion. Thus $nh = 0$ so $nw^2 = nw$.

Now $(1-w)^n = 1 - nw + w^2[\binom{n}{2} - \binom{n}{3}w + \dots] =$ $1 + w^2[-n + \binom{n}{2} - \binom{n}{3}w + \dots] = 1 + w^2 p$ for some $p:B \to B$. Since $n \geq 2$ and $(1-w)^2 w^2 = 0$, we have $(1-w)^n w^2 = 0$ so $(1-w)^n(1-w)^n = (1-w)^n$. Let $f = 1 - (1-w)^n$. Then $f^2 = f$ and $fp = p - (1-w)^n p = p - p(1-e)^n = p - p(1-e) = pe$. Similarly $gq = qf$. Clearly $e \neq 0,1$ implies $f \neq 0,1$ and the argument in the proof of the first part shows that $y \neq 0,1$. Thus the Lemma is proved.

We shall now show that an idempotent on K extends to one on SB. In order to do this we need to consider the following: let $R = \mathrm{End}(SB) \cong \mathrm{End}(B)$. Let $T \subset R$ be the torsion subgroup of R. Let $A = R/T$. Let $G:\mathcal{S} \to \mathfrak{m}_A$ be the functor assigning to an object Y the right A-module $G(Y) = \mathcal{S}(Y,SB)/\mathcal{S}(Y,SB)\cdot T$.

We shall prove later the following

<u>Lemma 4.49</u>: Let $q:B \to C$ be an epimorphism in $\mathfrak{m}_A$ where B is free and C is finite and p-torsion. If $g:C \to C$ is an idempotent, then there is an idempotent $f:B \to B$ with $gq = qf$.

Now given an idempotent $e:K \to K$, we get an idempotent $G(e):G(K) \to G(K)$. $G(K)$ is p-primary, since K is, $G(SB) \to G(K)$ is an epimorphism, and $G(SB) = A$ is free. Thus there is an idempotent $\bar{f}:G(SB) \to G(SB)$ a map of A-modules. Let $\tilde{f}:SB \to SB$ be such that $[\tilde{f}] = \bar{f}(1_{SB}) \in G(SB)$. Then $G(\tilde{f}) = \bar{f}$. Since $\bar{f}$ is an idempotent $G(\tilde{f}^n - \tilde{f}) = \bar{f}^n - \bar{f} = 0$. Thus $\tilde{f}^n = \tilde{f} + t_n$ where $t_n \in T$. But T is finite so by the argument used in Prop. 4.17, some power of $\tilde{f}$ is its own square; let $\hat{f} = \tilde{f}^n$ such that $\hat{f}^2 = \hat{f}$. Furthermore $G(\hat{f}) = \bar{f}$. Now $p:K \to SB$ is such that $G(\hat{f}p - pe) = 0$

Thus $\hat{f}p - pe \in \mathcal{S}(K,SB)\cdot T$. That is, $\hat{f}p - pe = \Sigma\, \alpha_i \beta_i$, $\beta_i \in \mathcal{S}(K,SB)$, $\alpha_i \in T$. Since SB is injective

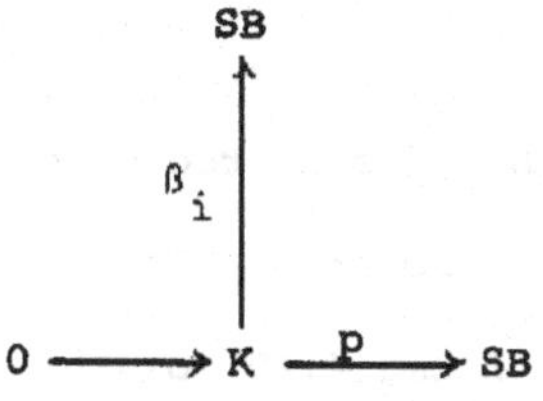

can be extended to

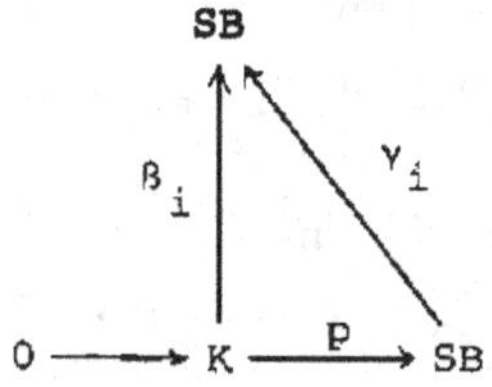

Thus $\gamma_i p = \beta_i$ so that $\hat{f}p - pe = \Sigma\, \alpha_i \gamma_i p$. Let $f' = \hat{f} - \Sigma\, \alpha_i \gamma_i$. Then $f' = \hat{f} + t$, $t \in T$ and $f'p = pe$. But f' is not necessarily an idempotent. However, $(f')^n = \hat{f}^n + nt\hat{f}^{n-1} + \ldots =$ $\hat{f} + t[n\hat{f} + \ldots] = \hat{f} + s_n$ where $s_n \in T$. We apply the same argument as before and set $f = (f')^n$ where $f^2 = f$. Then $fp = pe^n = pe$ and since $e \neq 0,1$, $f \neq 0,1$. Thus f is the desired idempotent and there remains only the proof of Lemma 4.49.

Let us call the statement of the lemma, S_A because it is a statement about a certain ring A. First we observe that $S_{A_1}, S_{A_2}, \ldots, S_{A_n}$ imply $S_{A_1 \oplus \ldots \oplus A_n}$: let e_i be the idempotent "projection onto A_i." Given the A-module epimorphism $q: B \to C$ with B free and C p-torsion, we get the A_i-module epimorphisms $q_i: Be_i \to Ce_i$ with Ce_i p-torsion and Be_i A_i-free. An idempotent

$g:C \to C$ yields $g_i:C_i \to C_i$ and hence may be lifted to $f_i:Be_i \to Be_i$ an idempotent by S_{A_i}. Then $f = f_1 \oplus \ldots \oplus f_n$: $B \cong Be_1 \oplus \ldots \oplus Be_n \to Ce_1 \oplus \ldots \oplus Ce_n \cong C$ is defined and is an idempotent covering g.

In our case $A = G(SB)$ is a product of integral matrix rings since SB is a bouquet of spheres. Thus by the previous argument, it will suffice to prove Lemma 4.49 for the case of A an integral matrix ring. For convenience, we use left-modules instead of right.

Let $A = \mathrm{End}(Z^n)$, the ring of integral $(n \times n)$ matrices. Let $\varphi: {}_Z\mathfrak{m} \to {}_A\mathfrak{m}$ be the functor $\mathrm{Hom}(Z^n, \ )$. First observe that $\mathrm{Hom}_A(Z^n,Z^n) \cong Z$: if $f:Z^n \to Z^n$ is an A-module map, it is determined by $f(1,0,\ldots,0) = (a_1,a_2,\ldots,a_n)$: Let $x \in A$ be given by $x(x_1,\ldots,x_n) = (\sum_{i=1}^n x_i, \sum_{i=2}^n x_i,\ldots,x_n)$. Then $(\sum_{i=1}^n a_i, \sum_{i=2}^n a_i,\ldots,a_n) = xf(1,0,\ldots,0) = fx(1,0,\ldots) = f(1,0,0,\ldots) = (a_1,\ldots,a_n)$ so $a_1 = \sum_{i=1}^n a_i$ so $a_2 = \sum_{i=2}^n a_i = 0$ and $a_3 = \sum_{i=3}^n a_i = 0,\ldots$. Thus $f(1,0,\ldots) = (a_1,0,0,0)$ so f represents multiplication by a_1; hence is an element of Z. We can state this by saying $\varphi_{Z,Z}:\mathrm{Hom}_Z(Z,Z) \to \mathrm{Hom}_A(\varphi(Z),\varphi(Z))$ is an isomorphism. This argument clearly extends to direct sums - i.e. $\varphi_{G,G'}$ is an isomorphism if G,G' are free. Furthermore observe that any free A-module is isomorphic to something in the image of φ since $A = \varphi(Z^n)$. Thus for any G and K with G free $\varphi_{G,K}$ is an isomorphism by the right exactness of $\mathrm{Hom}_A(\varphi(G), -)$ and then, from the right exactness of $\mathrm{Hom}_A(-,\varphi(K))$ we have $\varphi_{L,K}$ an isomorphism for any L and K. Finally if we consider an arbitrary A-module M, it is isomorphic to a quotient $F_i/f(F_2)$ where $f:F_2 \to F_1$ is a map of free A-modules. Then f is equivalent to $\varphi(f'):\varphi(F_2') \to \varphi(F_1')$ where

$f':F_2' \to F_1'$ is a map of free Z-modules. Then from the exactness of φ, $\varphi(F_2'/f'(F_1')) \cong F_1/f(F_2) \cong M$.

Thus any map in ${}_A\mathfrak{m}$ is isomorphic to one in the image of φ. Hence the statement S_A follows from the statement S_Z, so our problem is reduced to showing that S_Z is true. (Incidentally, the above equivalence of categories is more general: if R is any ring and $A = \text{End}(R^n)$ then $\varphi: {}_R\mathfrak{m} \to {}_A\mathfrak{m}$ is an equivalence.)

Theorem 4.50: Let F be a free abelian group, T a group with exponent p^N and $q:F \to T$ an epimorphism. If $T = T_1 \oplus T_2$ where T_1 is cyclic, then there is a splitting $F = F_1 \oplus F_2$ where $q(F_i) = T_i$ and F_1 is cyclic.

Proof: Let $a \in F$ be such that $q(a)$ generates T_1. Let $F_1 = \{x \in F \mid \exists m,n \in Z \ni: p \nmid m \text{ with } mx = na\}$. We claim that F_1 is a cyclic summand of F. First observe that $F_1 \otimes \mathbb{Q}$ is a one-dimensional $\mathbb{Q}$-vector space since it is clearly generated by $a \otimes 1$. Thus F_1 is a free abelian group of rank 1. To show that it is a summand it will suffice to show that F_1 is pure - i.e. $mx \in F$, for some integer m implies $x \in F_1$. But this is trivially true if $p \nmid m$ so it will suffice to show that $px \in F_1$ implies $x \in F_1$. So assume that $px = na$. Let $q(x) = tqa + \varphi$ where $\varphi \in T_2$. Then $nq(a) = q(na) = q(px) = ptq(a) + p\varphi$. By the splitting, then, $p\varphi = 0$ in T_2, $(pt - n)q(a) = 0$. Since $q(a)$ is of order a power of p, p divides $pt - n$, hence p divides n. Thus $1 \cdot x = (\frac{n}{p})a$ so $x \in F_1$. Thus F_1 is a cyclic summand of F.

Let α be a generator of F_1. Then $a = m_1\alpha$ for some integer m_1 and $m_2\alpha = na$ for integers m_2 and n where $p \nmid m_2$. Thus we can find integers s,t such that $sm_1 + tp^N = 1$. Then $sa = sm_1\alpha = \alpha - tp^N\alpha$ so $sa + tp^N\alpha = \alpha$. But then $q(\alpha) = sq(a) \in T_1$ so $q(F_1) \subset T_1$, but $q(a) = m_1q(\alpha)$ generates T_1 so $q(F_1) = T_1$. Extend α to a basis $\{\alpha\} \cup \{\alpha_i\}$. $q(\alpha_i) = s_iq(\alpha) + \varphi_i$ for integers s_i, and $\varphi_i \in T_2$. Then let F_2 be the summand of F generated by $\{\alpha_i - s_i\alpha\}$. Clearly $F = F_1 \oplus F_2$ and $q(F_2) \subset T_2$ since $q(\alpha_i - s_i\alpha) = \varphi_i$.

<u>Corollary 4.51</u>: Let F be a free abelian group, T a finite p-torsion group, $q:F \to T$ an epimorphism. Then if $e:T \to T$ is an idempotent, there exists an idempotent $f:F \to F$ with $eq = qf$.

<u>Proof</u>: $T = T_1 \oplus T_2$ where $T_1 = \text{im } e$, $T_2 = \text{im}(1-e)$. We can write $T_1 = \bigoplus_{i=1}^{n} C_i$ where C_i is cyclic. Then applying the theorem inductively we write $F = \bigoplus_{i=1}^{n} F_i \oplus \bar{F}$ where $q(F_i) = C_i$ and $q(\bar{F}) = T_2$. Then let f be the projection onto $\bigoplus_{i=1}^{n} F_i$.

This completes the proof of Lemma 4.49, hence Theorem 4.47 (that $\text{End}_{\mathcal{S}_p}$ (p-primary indecomposable) is local) and thus the main theorem, 4.44, that $K_o(\mathcal{S})$ is free.

CHAPTER 5. SPECTRAL SEQUENCES AND CALCULATION OF STABLE HOMOTOPY

5.1 Construction of the Spectral Sequence

In this chapter we shall make use of the construction of spectral sequences from exact couples. We recall briefly that an exact couple $(D,E;i,j,k)$ is a pair of abelian groups D and E and an exact triangle

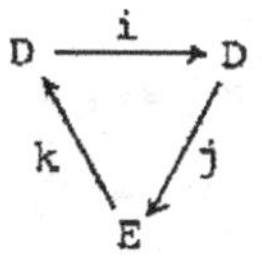

Given an exact couple, we get the derived couple $(D',E';i',j',k')$ where $D' = \text{im } i$ and $E' = \dfrac{\ker jk}{\text{im } jk}$; $i' = i|D'$, k' is induced by k and $j'(i(d)) = j(d) + \text{im } jk$ for $i(d') \in D'$. The derived couple is an exact couple and, leads to a new derived couple. $(D'',E'';i'',j'',k'')$. The process continues. Let the n-th derived couple be $(D^{(n)},E^{(n)};i^{(n)},j^{(n)},k^{(n)})$. Let $\partial^n = j^{(n)}k^{(n)}$. Then $E^{(n+1)}$ is the homology of $E^{(n)}$ with respect to $\partial^{(n)}$. In the usual cases, $E^{(n)} = E^{(n)}_{**}$ is a bigraded group and for each (s,t) $\partial^{(n)}|E^{(n)}_{s,t}$ is zero for sufficiently large n . Thus the $E^{(n)}_{s,t}$ stabilize for each (s,t) . These terms are written as $E^{\infty}_{s,t}$. E^{∞}_{**} is a bigraded group related to a filtration of some group we are interested in.

For example, let $k_*:\mathcal{C}^2 \to \mathcal{G}$ be a functor assigning to each space and subspace pair (X,A) a sequence of abelian groups $k_n(X,A)$ with $k_n(X,\emptyset)=k_n(X)$ such that there is a natural long exact sequence

$$\dots \to k_n(A) \to k_n(X) \to k_n(X,A) \to k_{n-1}(A) \to k_{n-1}(X) \to \dots .$$

(For example, $k_* = \pi_*$ or $k_* = H_*(\ ;\underline{A})$, the unreduced generalized homology theory with coefficients in the spectrum $\underline{A}$.)

Then let $* = W_{-1} \subset W_0 \subset W_1 \subset \ldots \subset W_n \subset W_{n+1} \subset \ldots$ be an increasing sequence of subspaces of $W = \bigcup_{i=0}^{\infty} W_i$. Then there is a bigraded exact couple with $E_{s,t} = k_{s+t}(W_s, W_{s-1})$ and $D_{s,t} = k_{s+t}(W_s)$. Assume $k_i(W_s, W_{s-1}) = 0$ for $i < s$. Then this leads to a spectral sequence

Theorem 5.1: There is a spectral sequence $\{E^r_{**}, d^r\}$ with $E^1_{s,t} = k_{s+t}(W_s, W_{s-1})$ and $d^1: E^1_{s,t} \to E^1_{s-1,t}$ is the composite $k_{s+t}(W_s, W_{s-1}) \to k_{s+t-1}(W_{s-1}) \to k_{s+t-1}(W_{s-1}, W_{s-2})$. In general, $d^r: E^r_{s,t} \to E^r_{s-r,t+r-1}$. Letting $F_{s,t} = \operatorname{im}(k_{s+t}(W_s) \to k_{s+t}(W))$, $E^{\infty}_{s,t} \cong F_{s,t}/F_{s-1,t+1}$.

Full details and discussion can be found in Spanier, Chapter 9.

One immediate use is the following: let X be a CW complex with skeleta X^n . Then form the spectral sequence for $X^1 \subset \ldots \subset X^n \subset X^{n+1} \subset \ldots \subset X$ and H_* . Then

$$E^1_{p,q} = H_{p+q}(X^p, X^{p-1}) = \begin{cases} 0 & q \neq 0 \\ C_p & q = 0 \end{cases}$$

where C_p is the free abelian group on the set of p-cells, J_p . Then $d^1: E^1_{p,0} \to E^1_{p-1,0}$ is a map $\partial_p: C_p \to C_{p-1}$ and clearly $E^2 = E^{\infty}$. Thus $H_*(C_*, \partial_*) \cong H_*(X)$. Observe also that $C_p \cong \tilde{H}_p(X^p/X^{p-1}) \cong \pi_p(X^p/X^{p-1})$ (made abelian for $p = 1$) and ∂_p corresponds to the maps induced by the composite $X^p/X^{p-1} \to SX^{p-1} \to S(X^{p-1}/X^{p-2})$. Also recall that $X^p/X^{p-1} \simeq \bigvee_{J_p} S^p$.

Now let X be a CW complex and Y an $(m-1)$-connected space.

Let $W = Y \wedge X$, $W_r = Y \wedge X^r$. Applying Theorem 5.1 for $k_* = \pi_*$ yields a spectral sequence with $E^1_{p,q} = \pi_{p+q}(Y \wedge X^p, Y \wedge X^{p-1})$, and d^1 the boundary map of the triple $(Y \wedge X^p, Y \wedge X^{p-1}, Y \wedge X^{p-2})$.

Now W_{p-1} is $(m-1)$-connected and (W_p, W_{p-1}) is $(m+p-1)$-connected thus $\pi_{p+q}(W_p, W_{p-1}) \to \pi_{p+q}(W_p/W_{p-1})$ is an isomorphism for $p+q < 2m+p-1$ by Theorem 1.8. Now the map $W_p/W_{p-1} = Y \wedge X^p/Y \wedge X^{p-1} \to Y \wedge (X^p/X^{p-1})$ is a weak homotopy equivalence (a homeomorphism for Y sufficiently nice). Thus for $t < 2m - 1$, $E^1_{p,q} \cong \pi_{p+q}(Y \wedge (X^p/X^{p-1}))$. But $X^p/X^{p-1} \cong \bigvee_{J_p} S^p$. By Theorem 1.8, $\pi_{p+q}(Y \wedge (X^p/X^{p-1})) \cong$

$\bigoplus_{J_p} \pi_{p+q}(Y \wedge S^p) \cong C^p \otimes \pi_{p+q}(S^p Y)$ for $p + q + 1 < 2(p + m)$, in particular for $q < 2m - 1$.

Furthermore, $\pi_{p+q}(S^p Y) \cong \pi_q(Y)$ for $q < 2m - 1$ by Theorem 1.9. Thus for $q < 2m - 1$, $E^1_{p,q} \cong C_p \otimes \pi_q(Y)$ and $d^1 : C_p \otimes \pi_q(Y) \to C_{p-1} \otimes \pi_q(Y)$ is clearly $\partial_p \otimes 1_{\pi_q(Y)}$. Thus $E^2_{p,q} \cong H_p(X; \pi_q(Y))$ for $q < 2m - 1$.

Furthermore $E^\infty_{p,q} \cong F_{p,q}/F_{p-1,q+1}$ where $F_{p,q} = \operatorname{im}(\pi_{p+q}(Y \wedge X^p) \to \pi_{p+q}(Y \wedge X))$. But $(Y \wedge X, Y \wedge X^p)$ is $(p+m)$-connected so $F_{p,q} = \pi_{p+q}(Y \wedge X)$ for $q \leq m$. Thus E^∞_{**} gives a graded group associated to all of $\pi_*(X \wedge Y)$. This yields the Whitehead spectral sequence (G. W. Whitehead [1]):

Theorem 5.2: If X is a CW space and Y is an $(m-1)$-connected space, then there is a spectral sequence converging to $\pi_*(X \wedge Y)$ with $E^2_{s,t} \cong H_s(X; \pi_t(Y))$ for $t \leq 2m - 2$.

Remark: Let $Y = K(G,m)$ and compare this to Theorem 3.13.

Theorem 5.3: For any CW space X and convergent spectrum $\underline{A}$, there exists a spectral sequence natural in both variables, with $E^2_{s,t} = H_s(X;\pi_t(\underline{A}))$ converging to $H_*(X;\underline{A})$.

Proof: Let ${}^nE^r_{**}$ be the spectral sequence converging to $\pi_*(X \wedge A_n)$. Assuming (for convenience) that $\underline{A}$ is strongly convergent and A_n is (n-1)-connected, ${}^nE^2_{s,t+n} = H_s(X;\pi_{t+n}(A_n))$ for $t + n \leq 2n - 2$ (hence for $t \leq n - 2$). By the naturality of the spectral sequence, there exist maps ${}^nE^r_{s,t+n} \to {}^{n+1}E^r_{s,t+n+1}$, which are isomorphisms for n sufficiently large and r,t,s fixed. Then let $E^r_{s,t} = {}^nE^r_{s,t+n}$ for sufficiently large n, then $E^2_{s,t} \cong H_s(X;\pi_t(\underline{A}))$ and the spectral sequence converges to $\pi_*(X \wedge \underline{A}) = H_*(X;\underline{A})$.

We can pass immediately to the "more stable" case and replace X by a spectrum:

Corollary 5.4: If $\underline{B}$ is a spectrum and $\underline{A}$ a convergent spectrum then there is a natural spectral sequence with $E^2_{s,t} = H_s(\underline{B};\pi_t(\underline{A}))$ and converging to $H_*(\underline{B};\underline{A})$.

Note $H_i(\underline{B};\underline{A}) = \varinjlim_n H_{n+i}(B_n;\underline{A}) = \varinjlim_n \varinjlim_m \pi_{n+i+m}(B_n \wedge A_m) =$
$\varinjlim_m \varinjlim_n \pi_{n+i+m}(A_m \wedge B_n) = \varinjlim_m H_{m+i}(A_m;\underline{B}) = H_i(\underline{A};\underline{B})$.

Recall that $\tilde{H}_*(\ ;G) = H_*(\ ;\underline{K}(G))$. In particular then for any spectrum $\underline{A}$, $H_*(\underline{A};G) = H_*(\underline{A};\underline{K}(G)) = H_*(\underline{K}(G);\underline{A})$. Furthermore $\pi_*(\ ) = H_*(\ ;\underline{S})$ so $\pi_*(\underline{A}) = H_*(\underline{A};\underline{S})$. Thus we get

<u>Proposition 5.5</u>: Given a convergent spectrum $\underline{A}$ and an abelian group G there is a natural spectral sequence converging to $H_*(\underline{A};G)$ with $E^2_{s,t} = H_s(\underline{K}(G);\pi_t(\underline{A}))$. Given a spectrum $\underline{B}$ there is a natural spectral sequence converging to $\pi_*(\underline{B})$ with $E^2_{s,t} = H_s(\underline{B};\pi_t(\underline{S}))$.

We now consider a special case of both of the above: take $\underline{A} = \underline{S}$ and $\underline{B} = \underline{K}$. Then we get a spectral sequence converging to $H_*(\underline{S}) = \pi_*(\underline{K}) = \begin{cases} Z & \text{in degree } 0 \\ 0 & \text{elsewhere} \end{cases}$. $E^2_{**} = H_*(\underline{K};\pi_*(\underline{S}))$. This gives us some information about $\pi_*(\underline{S})$ in terms of $H_*(\underline{K})$. In fact $H_*(\underline{K})$ is a combinatorially defined ring and is known (<u>Cartan Seminar</u>).

For example Theorem 4.11 says that $H_i(\underline{K})$ is finite for $i > 0$. In addition $H_o(\underline{K}) = Z$ and ${}^pH_i(\underline{K}) = \begin{cases} 0 & 0 < i < 2p-2 \\ Z_p & i = 2p-2 \end{cases}$.

(${}^pG = \{\theta \in G | p^n\theta = 0$ for some $n\}$ is the p-primary part of G).

This gives us the following information about $\pi_*(\underline{S})$ (this gives an alternate proof of Cor. 4.12):

<u>Theorem 5.6</u>: $\pi_i(\underline{S})$ is finite if $i > 0$ and ${}^p\pi_i(\underline{S}) = 0$ for $0 < i < 2p-3$, ${}^p\pi_{2p-3}(\underline{S}) \cong Z_p$.

<u>Proof</u>: Assume that $\pi_i(\underline{S})$ is finite for $0 < i < r$. Then for $r > 0$ $E^\infty_{0,r} = 0$ so every element in $E^2_{0,r} \cong \pi_r(\underline{S})$ is the boundary of some element (or coset of elements) in some $E^2_{s,r+1-s}$ for $s \geq 2$; but $\sum_{s=2}^{r+1} E^2_{s,r+1-s} = \sum_{s=2}^{r+1} H_s(\underline{K};\pi_{r+1-s}(\underline{S}))$ is finite hence $\pi_r(\underline{S})$ is finite.

Next we look at a picture of the spectral sequence. We will put $E_{s,t}$ in column $(s+t)$ and row t. Thus the base (row 0) is $H_*(\underline{K})$, and the fibre (which runs diagonally) is $\pi_*(\underline{S})$. Since everything is finite in positive degrees the spectral sequence breaks up into its p-primary parts. Then we have

$2p-3$						π_{2p-3}	
$\vdots$							
2			π_2	0	$\cdots$	0	0
1		π_1	0	0	$\cdots$	0	0
	Z	0	0	$\cdot$	$\cdot\ \cdot$	0	H_{2p-2}
	0	1	2			$2p-3$	$2p-2$

where $\pi_i = {}^P\pi_i(\underline{S})$ and $H_i = {}^PH_i(\underline{K})$. From the picture we see that in order to get $E^\infty_{s,t} = 0$ for $(s,t) \neq (0,0)$ we must have $\pi_i = 0, \ 0 < i < 2p-3$ and $\pi_{2p-3} \cong H_{2p-2} \cong Z_p$ as claimed.

5.2 The Cohomology of $\underline{K}$ and $\underline{K}_p$

Our program from now on is to get as much information as possible about $\pi_*(\underline{S})$ by feeding in long-known results (which we shall not prove) about $H_*(\underline{K})$.

First we shall have to study the spectra $\underline{K} = \underline{K}(Z)$ and $\underline{K}_p = \underline{K}(Z_p)$. Throughout this discussion we shall let $\iota_n \in H^n(K(G,n);G)$ represent the fundamental class - i.e. the class of the identity in $[K(G,n),K(G,n)]$. From the Künneth formula, we have that for a ring R

$$H^n(X;R) \otimes_R H^m(Y;R)$$

is a summand of

$$H^{n+m}(X \wedge Y;R)$$

in a natural way. In particular then, $\iota_n \otimes \iota_m \in H^n(K(R,n);R) \otimes_R H^m(K(R,m);R)$ represents an element of $H^{n+m}(K(R,n) \wedge K(R,m);R)$, hence a class of maps $f_{n,m}:K(R,n) \wedge K(R,m) \to K(R,n+m)$. $f_{n,m}$ represents "cup product" in the following way: Let $x \in \tilde{H}^n(X;R)$, $y \in \tilde{H}^m(X;R)$. Pick maps $\alpha_x:X \to K(R,n)$, $\alpha_y:X \to K(R,m)$ representing x and y . Then we can define $(\alpha_x \wedge \alpha_y)\circ\Delta:X \to K(R,n) \wedge K(R,m)$, where $\Delta:X \to X \wedge X$ is the reduced diagonal. Then composing with $f_{n,m}$ we get $f_{n,m}\circ(\alpha_x \wedge \alpha_y)\circ\Delta:X \to K(R,n+m)$ representing a cohomology class in $\tilde{H}^{n+m}(X;R)$. This cohomology class is precisely $x \cup y$.

In cohomology we have the natural transformation "loop suspension" $\sigma:H^{n+1}(-;G) \to H^n(\Omega-;G)$ given by $[-,K(G,n+1)] \xrightarrow{\Omega} [\Omega-,K(G,n)]$.

Clearly $\sigma(\iota_n) = \iota_{n-1}$. Thus the diagram

$$\begin{array}{ccccc}
K(R,n) \wedge K(R,m) & & \xrightarrow{\quad f_{n,m} \quad} & & K(R,n+m) \\
\Big\downarrow \wr & & & & \Big\downarrow \wr \\
(\Omega K(R,n+1)) \wedge K(R,m) & \xrightarrow{\theta} & \Omega(K(R,n+1) \wedge K(R,m)) & \xrightarrow{\Omega f_{n+1,m}} & \Omega K(R,n+m+1)
\end{array}$$

commutes, (where $\theta(\omega \wedge x)(t) = \omega(t) \wedge x$) if the $f_{n,m}$ are chosen properly.

By a similar diagram for $K(R,m) \simeq \Omega K(R,m+1)$, then, we can show that $\underline{K}(R)$ is an example of the following:

<u>Definition:</u> A ring spectrum $\underline{R}$ is a spectrum $\{R_n\}$ together with maps $\alpha_{n,m} : R_n \wedge R_m \to R_{n+m}$ such that

$$\begin{array}{ccccc}
 & (SR_n) \wedge R_m & \longrightarrow & R_{n+1} \wedge R_m & \\
{}^{T_1}\nearrow & & & & \searrow^{f_{n+1,m}} \\
S(R_n \wedge R_m) & \xrightarrow{Sf_{n,m}} & SR_{n+m} & \longrightarrow & R_{n+m+1} \\
{}_{T_2}\searrow & & & & \nearrow_{f_{n,m+1}} \\
 & R_n \wedge (SR_m) & \longrightarrow & R_n \wedge R_{m+1} &
\end{array}$$

commutes where the maps T_1 and T_2 are twists with a sign.

If $\underline{R}$ is a ring spectrum then $\pi_*(\underline{R})$ and $H_*(\underline{R})$ are graded rings; i.e. there are maps $\pi_n(\underline{R}) \otimes \pi_m(\underline{R}) \to \pi_{n+m}(\underline{R})$ and $H_n(\underline{R}) \otimes H_m(\underline{R}) \to H_{n+m}(\underline{R})$. Note that these may be non-associative. Beside the ring spectrum $\underline{K}(R)$ for a ring R , we observe that since $S^n \wedge S^m = S^{n+m}$, $\underline{S}$ is a ring spectrum.

Now we can ask about the ring structure of $H_*(\underline{K}(R);R')$, R,R'

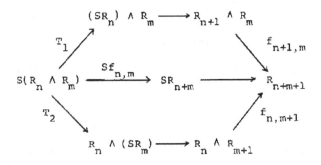

rings, and about the ring structure of $\pi_*(\underline{S})$. In particular we will look at the graded ring $G_* = H_*(\underline{K}_p;Z_p)$.

From Theorem 4.11 it follows that G_* is of finite type (that is, it is finitely generated in each degree). Then we can also look at $G^* = \mathrm{Hom}(G_*,Z_p)$ which is $H^*(\underline{K}_p;Z_p) \cong [\underline{K}_p,\underline{K}_p]^*$. But this latter group has a ring structure defined by composition of maps, (and of course it also represents all stable cohomology operations; i.e. natural transformations $H^n(\ ;Z_p) \to H^{n+*}(\ ;Z_p)$ commuting with suspension).

Thus G_* and its dual G^* (called the mod p Steenrod Algebra) both have ring (or rather Z_p-algebra) structures.

Thus we have maps

$$\varphi: G^* \otimes G^* \to G^*$$

$$\psi^*: G_* \otimes G_* \to G_*$$

Proposition 5.7: If A and B are graded vector spaces positive and of finite type (i.e. finitely generated in each degree and 0 in negative degrees) over a field k , then letting $A^* = \mathrm{Hom}_k(A,k)$, we have $A^{**} \cong A$ and $(A \otimes B)^* \cong A^* \otimes B^*$.

Proof: A^* in degree n is $A^n = \mathrm{Hom}_k(A_n,k)$ so $A^{**} \cong A$ is equivalent to the usual statement on finitely generated vector spaces (i.e. the existence of dual bases). By the same token $(A \otimes B)_n = \sum_{i=0}^{n} A_i \otimes B_{n-i}$ is finitely generated so the proof reduces to the usual statement in the ungraded case for finitely generated vector spaces. This is very easy: $A^* \otimes B^* \subset (A \otimes B)^*$ in a natural way; but both have the same (finite) dimension, hence they are equal.

Getting back to the maps φ and ψ^*, we can now take their duals $\varphi^* = \mathrm{Hom}(\varphi, Z_p)$ and $\psi = \mathrm{Hom}(\psi^*, Z_p)$. Composing with the isomorphisms of Proposition 5.7 gives

$$\varphi^* : \mathcal{G}_* \to \mathcal{G}_* \otimes \mathcal{G}_*$$

$$\psi : \mathcal{G}^* \to \mathcal{G}^* \otimes \mathcal{G}^* .$$

Then both $\mathcal{G}^*$ and $\mathcal{G}_*$ have two operations - a multiplication and a "comultiplication" or diagonal. In $\mathcal{G}_*$, $\psi^*(1 \otimes a) = \psi^*(a \otimes 1) = a$ so in $\mathcal{G}^*$, $\psi(a^*)$ contains the terms $1 \otimes a^*$ and $a^* \otimes 1$. More precisely, the diagram

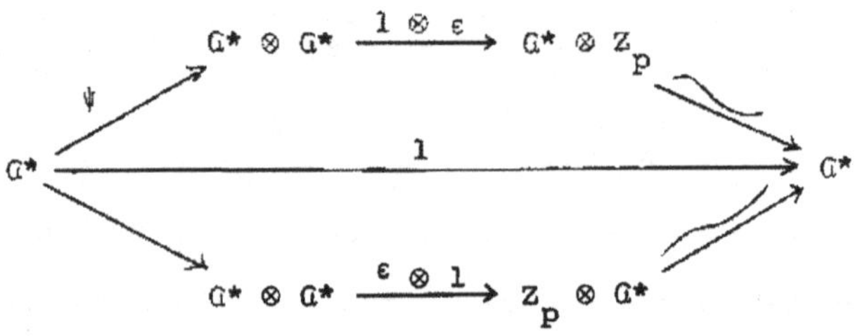

commutes, where ε is the augmentation, $\varepsilon|\mathcal{G}^n = 0$ for $n > 0$ and $\varepsilon|\mathcal{G}^0$ is the isomorphism $\mathcal{G}^0 \cong Z_p$.

Recall that ψ comes from the cup product map $\underline{K}_p \wedge \underline{K}_p \to \underline{K}_p$. We note the following connection: Let $T \in \mathcal{G}^*$ be such that $\psi(T) = \sum_i T_i' \otimes T_i''$. Then for any $x, y \in H^*(X; Z_p)$;,

$$T(x \cup y) = \sum_i (-1)^{|T_i''||x|} T_i'(x) \cup T_i''(y)$$

where we consider elements of $\mathcal{C}^*$ as cohomology operations and $|a|$ = degree of a. Then let $S \in \mathcal{C}^*$ such that $\psi(S) = \sum_j S'_j \otimes S''_j$.

Then $(ST)(x \cup y) = S(T(x \cup y)) = S(\sum_i (-1)^{|T''_i||x|} T'_i(x) \cup T''_i(y)) =$

$\sum_{i,j} (-1)^{|T''_i||x|} (-1)^{|S''_j||T'_i(x)|} S'_j T'_i(X) \cup S''_j T''_i(y)$. This sign is the same as $(-1)^{|S''_j T''_i||x|} \cdot (-1)^{|S''_j||T'_i|}$. Thus $\psi(ST) =$

$\sum_{i,j} (-1)^{|S''_j||T'_i|} S'_j T'_i \otimes S''_j T''_i$. But in the case of the tensor product of graded algebras, this latter term is precisely,

$(\sum S'_j \otimes S''_j) \cdot (\sum T'_i \otimes T''_i) = \psi(S)\psi(T)$. Thus $\psi(ST) = \psi(S)\psi(T)$.

Thus $\psi: \mathcal{C}^* \to \mathcal{C}^* \otimes \mathcal{C}^*$ is a homomorphism of algebras; another way of putting it is that the diagram

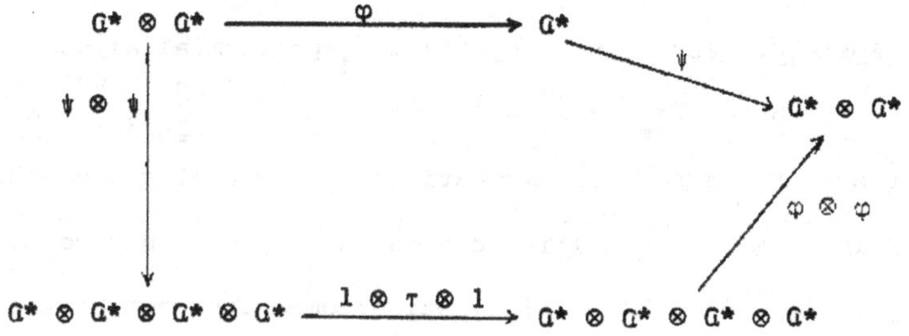

commutes, where $\tau(a \otimes b) = (-1)^{|b||a|} b \otimes a$.

Finally we observe that $1 \in \mathcal{C}^0$ is the multiplicative identity; i.e.

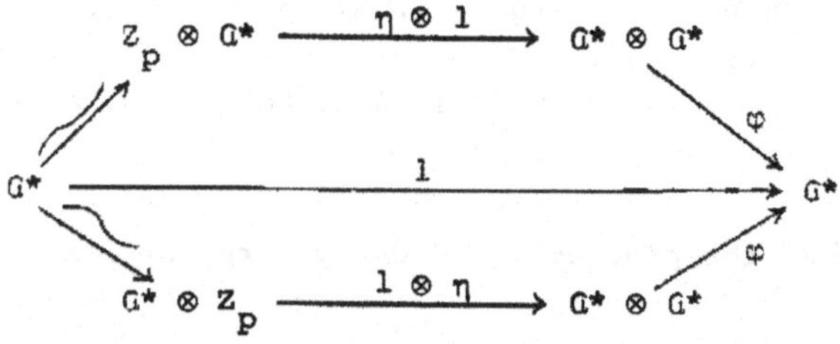

is commutative where $\eta: Z_p \xrightarrow{\simeq} \mathcal{G}^0$ is the unit map.

These three diagrams define

<u>Definition</u>: A Hopf algebra $(A,\epsilon,\eta,\varphi,\psi)$ over a field k is a graded k-vector space A with maps $\epsilon: A \to k$, $\eta: k \to A$, $\varphi: A \otimes A \to A$, $\psi: A \to A \otimes A$ such that the three appropriate diagrams (i.e. equivalent to the above) commute.

If $(A,\epsilon,\eta,\varphi,\psi)$ is a Hopf algebra, positive and of finite type, then $(A^*,\eta^*,\epsilon^*,\psi^*,\varphi^*)$ is a Hopf algebra. Furthermore, then, the double dual is isomorphic to the original. Thus given a Hopf algebra, it can be described entirely in terms of its dual.

In particular, we shall get our hands on $\mathcal{G}^* = [\underline{K}_p,\underline{K}_p]^*$ by describing instead $\mathcal{G}_* = H_*(\underline{K}_p;Z_p)$.

All the information about the Steenrod Algebra can be found in <u>Steenrod-Epstein</u>. For $p = 2$, $\mathcal{G}_*$ is a Z_2-polynomial algebra $P(\xi_1,\xi_2,\ldots)$ where $|\xi_n| = 2^n - 1$ and $\varphi^*(\xi_n) = \sum_{i=0}^{n} \xi_i^{2^{n-i}} \otimes \xi_{n-i}$. Observe that φ^* is not "cocommutative" - we cannot switch the factors. Then φ - the multiplication in $\mathcal{G}^*$ - is not commutative. (Not surprising since the multiplication comes from composition.) The dual to ξ_1^r is $Sq^r \in \mathcal{G}^r$ and the Sq^r generate $\mathcal{G}^*$ as an algebra, but <u>not</u> freely: if $a < 2b$ then (Adem relation)

$Sq^a Sq^b = \sum_{t=0}^{[a/2]} \binom{b-t-1}{a-2t} Sq^{a+b-t} Sq^t$. From this it follows that a Z_2-basis is given by $\{Sq^{i_1} Sq^{i_2} \ldots Sq^{i_r} | i_j \geq 2i_{j-1}\}$. It turns out that $Sq^{2^{n-1}} Sq^{2^{n-2}} \ldots Sq^2 Sq^1$ is dual to ξ_n with respect to this basis.

For $p > 2$ the problem is slightly more complex. We have

$\mathcal{a}_* = P(\xi_1,\xi_2,\ldots) \otimes E(\tau_o,\tau_1,\ldots)$ a Z_p-polynomial algebra $\otimes$ a Z_p-exterior algebra where the degrees are given by $1 + |\xi_n| = |\tau_n| = 2p^n - 1$.

We have P^r dual to ξ_1^r , β dual to τ_o and the P^t and β generate $\mathcal{a}^*$ as an algebra. $\varphi^*(\xi_n) = \sum_{i=0}^{n} \xi_i^{p^{n-i}} \otimes \xi_{n-i}$, $\varphi^*(\tau_n) = \tau_n \otimes 1 + \sum_{i=0}^{n} \xi^{p^{n-i}} \otimes \tau_{n-i}$. I will not go into the $p > 2$ Adem relation. I do however, wish to point out that since $P^r \longleftrightarrow \xi_1^r$ and $\beta \longleftrightarrow \tau_o$ we have $\psi(P^r) = \sum_{i=0}^{r} P^i \otimes P^{r-i}$ and $\psi(\beta) = 1 \otimes \beta + \beta \otimes 1$.

There are two actions of $\mathcal{a}^*$ on $\mathcal{a}_*$ given as follows: if $x \in \mathcal{a}^*$ and $a \in \mathcal{a}_*$ then ax and $xa \in \mathcal{a}_*$ are defined by the formulas $\langle ax,c\rangle = \langle a,xc\rangle$ and $\langle xa,c\rangle = \langle a,cx\rangle$ for all $c \in \mathcal{a}^*$ where $\langle\ ,\ \rangle : \mathcal{a}_* \otimes \mathcal{a}^* \to Z_p$ is the pairing. Note that if $x \in \mathcal{a}^n$ and $a \in \mathcal{a}_m$ then $ax, xa \in \mathcal{a}_{m-n}$.

Let $\mathcal{a}^{**} = \Pi \mathcal{a}^i$, $\mathcal{a}_{**} = \Sigma \mathcal{a}_i$. Then $\mathcal{a}^{**}$ acts on $\mathcal{a}_{**}$. $\mathcal{a}^{**}$ is a Hopf algebra and $\mathcal{a}^{**} = \mathrm{Hom}(\mathcal{a}_{**},Z_p)$. In particular let $P = P^o + P^1 + P^2 + \ldots$. Then

$$\psi(P) = \sum_{i=0}^{\infty} \psi(P^i) = \sum_{i=0}^{\infty} \sum_{j=0}^{i} P^j \otimes P^{i-j} = P \otimes P .$$

Then if $a,b \in \mathcal{a}_*$, $c \in \mathcal{a}^*$ with $\psi(c) = \Sigma\, c_i' \otimes c_i''$, then
$\langle (ab)P,c\rangle = \langle ab,Pc\rangle = \langle a\otimes b,\psi(Pc)\rangle = \langle a\otimes b,(P \otimes P)\cdot\Sigma\, c_i' \otimes c_i''\rangle =$
$\Sigma\langle a,Pc_i'\rangle\cdot\langle b,Pc_i''\rangle = \Sigma\langle aP,c_i'\rangle \langle bP,c_i''\rangle =$
$\langle aP \otimes bP,\Sigma c_i' \otimes c_i''\rangle = \langle (aP)(bP),c\rangle$. Thus $(ab)P = (aP)(bP)$ and similarly $P(ab) = (Pa)(Pb)$. Furthermore β is a derivation. Thus to evaluate $\mathcal{a}^*$ on $\mathcal{a}_*$ it suffices to find

$$\xi_n P,\ P\xi_n,\ \tau_n P,\ P\tau_n,\ \beta\xi_n,\ \xi_n\beta,\ \beta\tau_n,\ \tau_n\beta .$$

<u>Calculations</u>: $\langle \xi_n P, c\rangle = \langle \xi_n, Pc\rangle = \langle \Sigma \xi_i^{p^{n-i}} \otimes \xi_{n-i}, P \otimes c\rangle =$

$\Sigma\langle \xi_i^{p^{n-i}}, P\rangle \langle \xi_{n-i}, c\rangle = \langle \xi_o^{p^n}, P^o\rangle \langle \xi_n, c\rangle + \langle \xi_1^{p^{n-1}}, P^{p^{n-1}}\rangle \langle \xi_{n-1}, c\rangle =$

$\langle \xi_n + \xi_{n-1}, c\rangle$. So $\xi_n P = \xi_n + \xi_{n-1}$.

$\langle P\xi_n, c\rangle = \langle \xi_n, cP\rangle = \langle \Sigma \xi_i^{p^{n-i}} \otimes \xi_{n-i}, c \otimes P\rangle = \Sigma\langle \xi_i^{p^{n-i}}, c\rangle \langle \xi_{n-i}, P\rangle =$

$\langle \xi_n, c\rangle \langle \xi_o, P^o\rangle + \langle \xi_{n-1}^p, c\rangle \langle \xi_1, P^1\rangle = \langle \xi_n + \xi_{n-1}^p, c\rangle$. So $P\xi_n =$

$\xi_n + \xi_{n-1}^p$.

The rest of the calculations are straightforward. Then we get

<u>Theorem 5.8</u>: $\xi_n P = \xi_n + \xi_{n-1}, \; P\xi_n = \xi_n + \xi_{n-1}^p ,$

$$\tau_n P = \tau_n + \tau_{n-1}, \; P\tau_n = \tau_n$$

$$\tau_k \beta = \delta_{o,k}, \quad \xi_k \beta = 0, \quad \beta\tau_k = \xi_k, \quad \beta\xi_k = 0 .$$

So, for example, if we wanted to calculate $P^p P^1(\tau_1 \xi_1^{p+1} \xi_2)$ we first calculate $P(\tau_1 \xi_1^{p+1} \xi_2) = (P\tau_1)(P\xi_1)^{p+1}(P\xi_2) = \tau_1(\xi_1 + 1)^{p+1}(\xi_2 + \xi_1^p) =$
$\tau_1(\xi_1^{p+1} + \xi_1^p + \xi_1 + 1)(\xi_2 + \xi_1^p) = \tau_1 \xi_1^{p+1} \xi_2 + \tau_1 \xi_1^p \xi_2 + \tau_1 \xi_1 \xi_2 + \tau_1 \xi_2 +$
$\tau_1 \xi_1^{2p+1} + \tau_1 \xi_1^{2p} + \tau_1 \xi_1^{p+1} + \tau_1 \xi_1^p$. So $P^1(\tau_1 \xi_1^{p+1} \xi_2) = \tau_1(\xi_1^p \xi_2 + \xi_1^{2p+1})$
as it is the term of the correct degree.

Next we find $P(\tau_1(\xi_1^p \xi_2 + \xi_1^{2p+1})) =$
$\tau_1((\xi_1^p + 1)(\xi_2 + \xi_1^p) + (\xi_1 + 1)^{2p+1}) =$
$\tau_1(\xi_1^p \xi_2 + \xi_2 + \xi_1^{2p} + \xi_1^p + \xi_1^{2p+1} + \xi_1^{2p} + 2\xi_1^{p+1} + 2\xi_1^p + \xi_1 + 1)$ so
$P^p(\tau_1(\xi_1^p \xi_2 + \xi_1^{2p+1})) = \tau_1(\xi_2 + 2\xi_1^{p+1})$. Thus $P^p P^1(\tau_1 \xi_1^{p+1} \xi_2) =$

$\tau_1(\xi_2 + 2\xi_1^{p+1})$.

Now we get back to the spectral sequence

$E^r(\underline{A};\underline{B}):H_*(\underline{A};\pi_*(\underline{B})) \Rightarrow H_*(\underline{A};\underline{B})$.

Let us assume that we have a spectrum $\underline{B}$ with every element in $\pi_*(\underline{B})$ having order p and $\pi_n(\underline{B})$ finitely generated for each n. Thus $H_*(\underline{A};\pi_*(\underline{B})) \cong H_*(\underline{A};Z_p) \otimes \pi_*(\underline{B})$ and $\mathrm{Hom}(H_*(\underline{A};\pi_*(\underline{B})),Z_p) \cong H^*(\underline{A};Z_p) \otimes \pi^*(\underline{B})$ (where $\pi^*(\underline{B}) = \mathrm{Hom}(\pi_*(\underline{B}),Z_p)$).

Taking $\mathrm{Hom}(\ ,Z_p)$ of the spectral sequence we get a new spectral sequence $H^*(\underline{A};Z_p) \otimes \pi^*(\underline{B}) \Rightarrow (H_*(\underline{A};\underline{B}))^*$. This spectral sequence is clearly natural in $\underline{A}$ holding $\underline{B}$ fixed and vice versa.

Let us examine a special case of this for $\underline{A} = \underline{K}_p$. Hence we have

$$\mathcal{G}^* \otimes \pi^*(\underline{B}) \Rightarrow (H_*(\underline{K}_p;\underline{B}))^* = (H_*(\underline{B};Z_p))^* \cong H^*(\underline{B};Z_p)$$

Consider an element $1 \otimes \theta$ along the fibre. Say $d_r(1 \otimes \theta) = 0$ for $r < n$, $d_n(1 \otimes \theta) = \Sigma T_{\theta,\varphi} \otimes \varphi \in \mathcal{G}^* \otimes \pi^*(\underline{B})$.

φ runs through basis of $\pi^*(\underline{B})$

Now let $\underline{A}$ be any spectrum and $x \in H^*(\underline{A};Z_p)$.

Let $f:\underline{A} \to \underline{K}_p$ represent x .

Then look at $x \otimes \theta \in E^r(\underline{A};\underline{B})$.

$d_r(x \otimes \theta) = d_r(f^*(1) \otimes \theta) = f^*d_r(1 \otimes \theta)$ for $r \leq n$. Thus $d_r(x \otimes \theta) = 0$ for $r < n$ and $d_n(x \otimes \theta) = \Sigma f^*(T_{\theta,\varphi}) \otimes \varphi$. But of course $f^*(T_{\theta,\varphi}) = T_{\theta,\varphi}(x)$ where $T_{\theta,\varphi} \in \mathcal{G}^*$ is the cohomology operation. So $d_n(x \otimes \theta) = \Sigma T_{\theta,\varphi}(x) \otimes \varphi$.

Dually what this says is: If in $E^r(\underline{K}_p;\underline{B})$ we have $1 \otimes \theta \in H_*(\underline{K}_p;Z_p) \otimes \pi_*(\underline{B})$ an n-boundary, but not an (n-1)-boundary and if $d_n(\Sigma x_i \otimes b_i)$ hits $1 \otimes \theta$ then there exist elements T_i of $\mathcal{G}^n$ such that $\Sigma x_i T_i = 1$ and for any $\underline{A}$ and any $y \in H_*(\underline{A};Z_p)$, $y \otimes a$

is not an (n-1)-boundary but is n-boundary if there are $y_i \in H_*(\underline{A};Z_p)$ such that $\Sigma y_i T_i = y$. In practice and especially in low dimensions this Σ will only be over 1 element: i.e. we will find

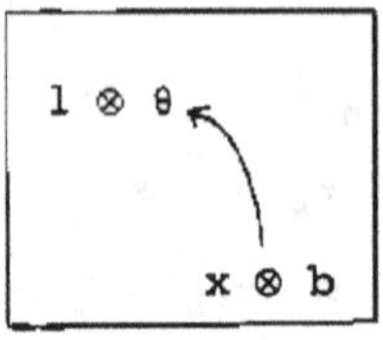

in $E^r(\underline{K}_p;\underline{B})$, then up to some indeterminacy we find an element $T \in G^*$ and we conclude

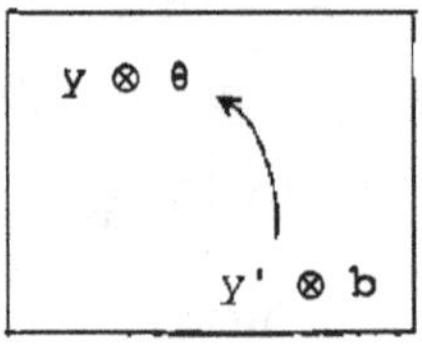

if and only if $y'T = y$ (up to any indeterminacy).

Since the left action of G^* on $H_*(\underline{K}_p;Z_p) = G_*$ can be thought of as coming from maps of spectra $\underline{K}_p \to K_p$, the naturality of the spectral sequence implies that this action commutes with the differentials.

There is one further technical property of the spectral sequence that we shall not prove:

If $\underline{A}$ and $\underline{B}$ are ring spectra, then the ring structures on $H_*(\underline{A})$ and $\pi_*(\underline{B})$ turn $H_*(\underline{A};\pi_*(\underline{B})) = E^2(\underline{A},\underline{B})$ into a ring and we can ask how d^2 acts on this. We find that (Dold or Kultze):

Theorem 5.9: d^2 is a derivation, so that E^3 is a ring; in general E^n is a ring and d^n a derivation; i.e. when the right hand side is defined, $d^n(xy) = d^n(x)y + (-1)^{n|x|}xd^n(y)$ for $x,y \in E^n$.

Observe that if $\underline{B}$ is as before (a p-primary spectrum) then $E^2(\underline{K}_p;\underline{B}) \cong G_* \otimes \pi_*(B) \cong E(\tau_o) \otimes A_* \otimes \pi_*(\underline{B})$ as an algebra, for any $A_* \cong P(\xi_1,\ldots) \otimes E(\tau_1,\ldots)$. Thus $E^2(\underline{K}_p,B) \cong [A_* \otimes \pi_*(B)] \oplus \tau_o[A_* \otimes \pi_*(\underline{B})]$.

We wish to prove that we can choose A_* correctly so that every differential respects this splitting.

Our choice for A_* is $H_*(\underline{K};Z_p)$. Now the exact sequence $0 \to Z \xrightarrow{p} Z \to Z_p \to 0$ yields the fibration $\underline{K} \xrightarrow{p} \underline{K} \xrightarrow{i_p} \underline{K}_p$ and the diagram

$$\begin{array}{ccc} \underline{K}_p & \xrightarrow{\sigma_p} & \underline{K} \\ & \searrow^{\beta} & \downarrow{\scriptstyle i_p} \\ & & \underline{K}_p \end{array}$$

commutes (where $\beta \in [\underline{K}_p,\underline{K}_p]^1$ is as before), thus

$$\begin{array}{ccccccc} 0 \to & H_n(\underline{K};Z_p) & \xrightarrow{i_p} & H_n(\underline{K}_p;Z_p) & \xrightarrow{\sigma_p} & H_{n-1}(\underline{K};Z_p) & \to 0 \\ & & \nearrow\!\!\wr & & & \downarrow{\scriptstyle i_p} & \\ & G_n & & \xrightarrow{\beta} & & G_{n-1} \cong H_{n-1}(\underline{K}_p;Z_p) & \end{array}$$

commutes. ($H_*(p;Z_p) = 0$, of course.)

So $A_n = H_n(\underline{K};Z_p) = \ker(\beta: G_n \to G_{n-1})$. Which action of β is this? Its dual in cohomology would be

$$\begin{array}{ccc} \beta^*: G_{n-1} & \longrightarrow & G_n \\ \| & & \| \\ H^{n-1}(\underline{K}_p;Z_p) & \longrightarrow & H^n(\underline{K}_p;Z_p) \end{array}$$

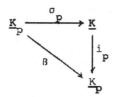

coming from a map of spectra $\underline{K}_p \to \underline{K}_p$ hence it would act on the right, so here $\beta : G_n \to G_{n-1}$ acts on the left. Recall $\beta\xi_n = 0$, $\beta\tau_n = \xi_n$. Thus $\ker \beta = P(\xi_1, \xi_2, \ldots) \otimes E(\tau_1 - \tau_o\xi_1, \tau_2 - \tau_o\xi_2, \ldots)$ Let $t_n = \tau_n - \tau_o\xi_n$, $n \geq 1$. Then we have $A_* =$ $P(\xi_1, \xi_2, \ldots) \otimes E(t_1, t_2, \ldots)$. Since we shall be using the t_n we can easily calculate the following: $t_n P = (\tau_n - \tau_o\xi_n)P =$ $(\tau_n + \tau_{n-1}) - \tau_o(\xi_n + \xi_{n-1}) = t_n + t_{n-1}$, $t_n\beta = -\xi_n$.

5.3 The Spectrum $\underline{L}_p$

Next we wish to find a spectrum $\underline{X}$ such that $\pi_*(\underline{X})$ has only p-torsion, but at the same time gives information concerning $\pi_*(\underline{S})$.

Let $p:\underline{S} \to \underline{S}$ of degree 0 represent multiplication by p . Let $\underline{L}_p$ be the cofibre. Then

$$\cdots \to \pi_n(\underline{S}) \xrightarrow{p} \pi_n(\underline{S}) \to \pi_n(\underline{L}_p) \to \pi_{n-1}(\underline{S}) \xrightarrow{p} \cdots$$

is exact. Since $\pi_n(\underline{S})/p\pi_n(\underline{S}) = \pi_n(\underline{S}) \otimes Z_p$ and $\ker p|\pi_{n-1}(\underline{S}) \cong \mathrm{Tor}(\pi_{n-1}(\underline{S}),Z_p)$, we have the exact sequence

$$0 \to \pi_n(\underline{S}) \otimes Z_p \to \pi_n(\underline{L}_p) \to \mathrm{Tor}(\pi_{n-1}(\underline{S}),Z_p) \to 0 .$$

So $\pi_n(\underline{L}_p)$ has only p- or perhaps p^2-torsion. But $\pi_o(\underline{L}_p) = Z_p$ and $\pi_1(\underline{L}_p) = \begin{cases} 0 & p \neq 2 \\ Z_2 & p = 2 \end{cases}$. From the exactness of

$$\begin{array}{ccccccccc}
[\underline{S},\underline{L}_p]_1 & \xrightarrow{p} & [\underline{S},\underline{L}_p]_1 & \to & [\underline{L}_p,\underline{L}_p]_o & \to & [\underline{S},\underline{L}_p]_o & \xrightarrow{p} & [\underline{S},\underline{L}_p]_o \\
 & & \wr\| & & & & \wr\| & & \downarrow\wr \\
 & & Z_2 \text{ or } 0 & & & & Z_p & & Z_p
\end{array}$$

we get: if $p > 2$ then $1:\underline{L}_p \to \underline{L}_p$ has order p .

If $f:X \to Y$ is any map, then the diagram

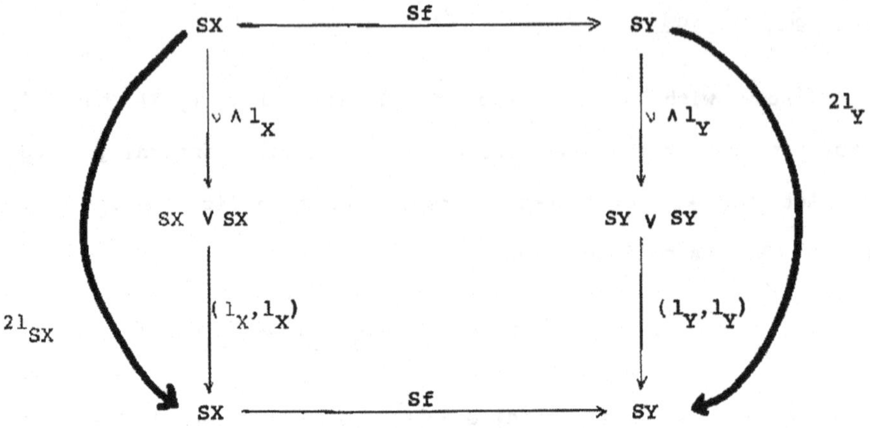

commutes where $\nu : S^1 \to S^1 \vee S^1$ is the usual pinch map. In particular this generalizes to say that the two actions of Z on $[SX,SY]$ are the same. Thus if 1_{SY} is of finite order n, then $n[SX,SY] = 0$.

Since the identity $\underline{L}_p \to \underline{L}_p$ is of order p for $p > 2$, it immediately follows that $\pi_*(\underline{L}_p)$ is a Z_p-module for $p > 2$. Thus for $p > 2$, $\pi_n(\underline{L}_p) \cong \pi_n(\underline{S}) \otimes Z_p \oplus \mathrm{Tor}(\pi_{n-1}(\underline{S}), Z_p)$ an isomorphism of groups.

For $p = 2$ we have

$$0 \to Z_2 \to [\underline{L}_2, \underline{L}_2]_o \to Z_2 \to 0$$

so that $[\underline{L}_2, \underline{L}_2]_o \cong Z_4$ or $Z_2 \oplus Z_2$. Thus twice the identity is either 0 or it maps to the generator of $\pi_1(\underline{L}_2)$.

At this point we recall the following (property 8 following the definition of mapping cone in Chapter 1):

If $f : X \to Y$ is any map and $f \wedge 1 : X \wedge Z \to Y \wedge Z$ is its smash with the identity of any space Z, then $C_{f \wedge 1} \cong C_f \wedge Z$.

This passes over to cofibrations of spectra. Since $\underline{S} \wedge \underline{L}_p \simeq \underline{L}_p$, smashing $\underline{L}_p$ with $\underline{S} \xrightarrow{p} \underline{S} \to \underline{L}_p$ yields the cofibration $\underline{L}_p \xrightarrow{p} \underline{L}_p \to \underline{L}_p \wedge \underline{L}_p$. But for p odd, $p1_{\underline{L}_p} \simeq *$ so that $\underline{L}_p \wedge \underline{L}_p \simeq \underline{L}_p \vee \underline{L}_p^1$. On the level of spaces, this says $M(Z_p,n) \wedge M(Z_p,m) \simeq M(Z_p,n+m) \vee M(Z_p,n+m+1)$.

The same will be true for $p = 2$ in case $1_{\underline{L}_2}$ is of order 2; i.e. if $[\underline{L}_2,\underline{L}_2] \simeq Z_2 \oplus Z_2$. Observe that in Z_2-cohomology

$$H^i(\underline{L}_2;Z_2) = \begin{cases} 0 & i \neq 0,1 \\ Z_2 & i = 0,1 \end{cases}$$

with (by the definition of the Bockstein, Sq^1) $Sq^1:H^0(\underline{L}_2;Z_2) \to H^1(\underline{L}_2;Z_2)$ an isomorphism. Now if $1_{\underline{L}_2}$ is of order 2 then $H^*(\underline{L}_2 \wedge \underline{L}_2;Z_2) \simeq H^*(\underline{L}_2 \vee \underline{L}_2^1;Z_2) \simeq H^*(\underline{L}_2;Z_2) \oplus H^{*-1}(\underline{L}_2;Z_2)$ as G^*-modules. (We know that $H^*(\underline{L}_2 \wedge \underline{L}_2;Z_2) \simeq H^*(\underline{L}_2;Z_2) \oplus H^{*-1}(\underline{L}_2;Z_2)$ as groups.) Let $0 \neq u \in H^0(\underline{L}_2;Z_2)$. Then for $u \otimes u \in H^0(\underline{L}_2 \wedge \underline{L}_2;Z_2) \simeq Z_2$, $Sq^2(u \otimes u) = 0$ by this splitting. The diagonal formula $\psi(Sq^2) = Sq^2 \otimes 1 + Sq^1 \otimes Sq^1 + 1 \otimes Sq^2$ tells us, however, that $Sq^2(u \otimes u) = Sq^1(u) \otimes Sq^1(u) \neq 0$. Thus $\underline{L}_2 \wedge \underline{L}_2 \not\simeq \underline{L}_2 \vee \underline{L}_2^1$ so $1_{\underline{L}_2}$ is of order 4. Thus $\pi_*(\underline{L}_2)$ may (in fact does) have 4-torsion as well as 2-torsion. For this reason, we now confine ourselves to $\underline{L}_p$ for p odd.

The splitting $M(Z_p,n) \wedge M(Z_p,m) \simeq M(Z_p,n+m) \vee M(Z_p,n+m+1)$ yields a map $f_{n,m}:M(Z_p,n) \wedge M(Z_p,m) \to M(Z_p,n+m)$ and it is immediate that this turns $\underline{L}_p$ into a ring spectrum for p odd. Let $\mu:\underline{L}_p \wedge \underline{L}_p \to \underline{L}_p$ represent the multiplication. We wish to check for the homotopy associativity of multiplication. That is, we want to know if the following diagram homotopy commutes (leading to $\pi_*(\underline{L}_p)$ being an associative ring).

$$\begin{array}{ccc}
\underline{L}_p \wedge \underline{L}_p \wedge \underline{L}_p & \xrightarrow{\mu \wedge 1} & \underline{L}_p \wedge \underline{L}_p \\
\downarrow {\scriptstyle 1 \wedge \mu} & & \downarrow {\scriptstyle \mu} \\
\underline{L}_p \wedge \underline{L}_p & \xrightarrow[\mu]{} & \underline{L}_p
\end{array}$$

Now

$$\begin{array}{ccc}
\underline{S} \wedge \underline{S} & \longrightarrow & \underline{S} \\
\downarrow {\scriptstyle i \wedge i} & & \downarrow {\scriptstyle i} \\
\underline{L}_p \wedge \underline{L}_p & \xrightarrow[\mu]{} & \underline{L}_p
\end{array}$$

commutes - which says that $i_*:\pi_*(\underline{S}) \to \pi_*(\underline{L}_p)$ is a ring map. Thus if we take $i \wedge i \wedge i:\underline{S} \wedge \underline{S} \wedge \underline{S} \to \underline{L}_p \wedge \underline{L}_p \wedge \underline{L}_p$ and compose with either $\mu\circ(1 \wedge \mu)$ or $\mu\circ(\mu \wedge 1)$ we still get i. Then if we could show that $[\underline{L}_p \wedge \underline{L}_p \wedge \underline{L}_p, \underline{L}_p]_o \simeq Z_p$ we would have homotopy commutativity of the first diagram - hence homotopy associativity of μ.

But $\underline{L}_p \wedge \underline{L}_p \wedge \underline{L}_p \simeq (\underline{L}_p \vee \underline{L}_p^1) \wedge \underline{L}_p \simeq (\underline{L}_p \wedge \underline{L}_p) \vee (\underline{L}_p^1 \wedge \underline{L}_p) \simeq \underline{L}_p \vee \underline{L}_p^1 \vee \underline{L}_p^1 \vee \underline{L}_p^2$ so

$[\underline{L}_p \wedge \underline{L}_p \wedge \underline{L}_p,\underline{L}_p]_o \simeq [\underline{L}_p,\underline{L}_p]_o \oplus ([\underline{L}_p,\underline{L}_p]_1)^2 \oplus [\underline{L}_p,\underline{L}_p]_2$.

Now from the cofibration $\underline{S} \xrightarrow{p} \underline{S} \to \underline{L}_p$ we get

$$\pi_n(\underline{L}_p) \xleftarrow{p} \pi_n(\underline{L}_p) \longleftarrow [\underline{L}_p,\underline{L}_p]_n \longleftarrow \pi_{n+1}(\underline{L}_p) \xleftarrow{p} \pi_{n+1}(\underline{L}_p) \ .$$

We noted before that $p\pi_*(\underline{L}_p) = 0$. Thus we have

$$0 \to \pi_{n+1}(\underline{L}_p) \to [\underline{L}_p,\underline{L}_p]_n \to \pi_n(\underline{L}_p) \to 0 \ .$$

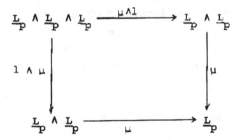

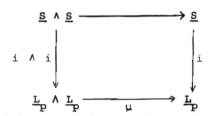

This also splits as Z_p-modules. Thus

$$[\underline{L}_p,\underline{L}_p]_n \cong \pi_{n+1}(\underline{L}_p) \oplus \pi_n(\underline{L}_p) \cong$$

$$(\pi_{n+1}(\underline{S}) \otimes Z_p) \oplus \mathrm{Tor}(\pi_n(\underline{S}),Z_p) \oplus (\pi_n(\underline{S}) \otimes Z_p) \oplus \mathrm{Tor}(\pi_{n-1}(\underline{S}),Z_p)$$

Thus based on what we know so far

$$[\underline{L}_p,\underline{L}_p]_n = \begin{cases} Z_p & n = -1,0,2p-4 \\ 0 & 1 \leq n \leq 2p-5 \end{cases}$$

Then for $p \geq 3$, $[\underline{L}_p,\underline{L}_p]_1 = 0$ and $[\underline{L}_p,\underline{L}_p]_2 \cong \begin{cases} Z_3 & p = 3 \\ 0 & p > 3 \end{cases}$.

Thus for $p > 3$ $\mu:\underline{L}_p \wedge \underline{L}_p \to L_p$ is homotopy associative, but for $p = 3$ it may fail. (In fact from a recent result of Toda (<u>Toda</u> [3]), it turns out that associativity <u>does</u> fail for $p = 3$.)

For $p = 3$, any failure of associativity will be due to the element $\alpha_1 \in \pi_3(\underline{S})$ of order 3 .

It is precisely this element that leads to the second Z_3-summand in $[\underline{L}_3 \wedge \underline{L}_3 \wedge \underline{L}_3,\underline{L}_3]_o$.

That is, the diagram

$$\begin{array}{ccc} \underline{L}_3 \wedge \underline{L}_3 \wedge \underline{L}_3 & \xrightarrow{\mu\circ(\mu\wedge 1)-\mu\circ(1\wedge\mu)} & \underline{L}_3 \\ \Big\downarrow{\scriptstyle \partial\wedge\partial\wedge\partial} & & \Big\uparrow{\scriptstyle i} \\ \underline{S} \wedge \underline{S} \wedge \underline{S} \cong \underline{S} & \xrightarrow{k\alpha_1} & \underline{S} \end{array}$$

homotopy commutes for some $k \in Z_3$. (Toda proves $k = \pm 1$ but we will not use this fact.)

$\partial : \underline{L}_p \to \underline{S}$ is σ_p. If we compose with $i_p : \underline{S} \to \underline{L}_p$ we get $\partial_p : \underline{L}_p \to \underline{L}_p$. We also write the induced maps as $\partial : \pi_n(\underline{L}_p) \to \pi_{n-1}(\underline{S})$, $\partial_p : \pi_n(\underline{L}_p) \to \pi_{n-1}(\underline{L}_p)$.

Let $f,g,h : \underline{S} \to \underline{L}_3$ represent $a,b,c \in \pi_*(\underline{L}_3)$. Then we have $a(bc) - (ab)c = [(\mu\circ(1\wedge\mu) - \mu\circ(\mu\wedge 1))\circ(f\wedge g\wedge h)] = [i\circ(k\alpha_1)\circ(\partial\wedge\partial\wedge\partial)\circ(f\wedge g\wedge h)] = ki_*\alpha_1[(\partial f)\wedge(\partial g)\wedge(\partial h)]$. Since i_* is a ring map this $= k\alpha(\partial_3 a)(\partial_3 b)(\partial_3 c)$ where $\alpha = i_*(\alpha_1) \in \pi_3(\underline{L}_3)$.

By following the exact sequences around, we see that if $\theta \in \pi_n(\underline{S})$ is of order p then we get (not uniquely) an element $\theta' \in \pi_{n+1}(\underline{L}_p)$ with $\partial(\theta') = \theta$, hence $\partial_p\theta' = i_*(\theta)$. If θ is not divisible by p then $i_*(\theta) \neq 0$, and we let $\theta \in \pi_n(\underline{L}_p)$ represent $i_*(\theta)$. Then $\theta \in \pi_n(\underline{S})$ of order p and not divisible by p leads to two Z_p-generators, $\theta' \in \pi_{n+1}(\underline{L}_p)$ and $\theta = \partial_p(\theta') \in \pi_n(\underline{L}_p)$.

The following relation may be checked easily: if $a,b \in \pi_*(\underline{L}_p)$, then
$$\partial_p(ab) = \partial_p(a)b + (-1)^{|a|}a\partial_p(b).$$
Finally since for $p > 2$, $[\underline{L}_p \wedge \underline{L}_p, \underline{L}_p]_o \cong [\underline{L}_p, \underline{L}_p]_o \oplus [\underline{L}_p, \underline{L}_p]_1 \cong Z_p$, we get that for the twist map $T : \underline{L}_p \wedge \underline{L}_p \to \underline{L}_p \wedge \underline{L}_p$, $T\circ\mu \sim \mu$ (using the same argument as for associativity). Thus μ is homotopy commutative, so $\pi_*(\underline{L}_p)$ is graded commutative.

<u>Theorem 5.10</u>: $\pi_*(\underline{L}_p)$ is a graded commutative ring. ∂_p is a derivation on $\pi_*(\underline{L}_p)$ and $\partial_p^2 = 0$. For $p > 3$, $\pi_*(\underline{L}_p)$ is associative and for $p = 3$, there is some fixed $k \in Z_3$, and $0 \neq \alpha \in \pi_3(\underline{L}_3)$ such that $a(bc) - (ab)c = k\alpha(\partial_3 a)(\partial_3 b)(\partial_3 c)$.

5.4 The Calculations

Now let us look at $E^r(\underline{K}_p;\underline{L}_p) \Rightarrow H_*(\underline{L}_p;Z_p)$ and $E^r(\underline{K};\underline{L}_p) \Rightarrow H_*(\underline{L}_p;Z)$ for $p \geq 3$. Recall that in pictures, $E^2_{s,t}$ will be in column $(s+t)$, row t. Thus d^r moves left one column, up $(r-1)$ rows. All blanks represent 0. All letters represent Z_p-generators.

A beginning picture is

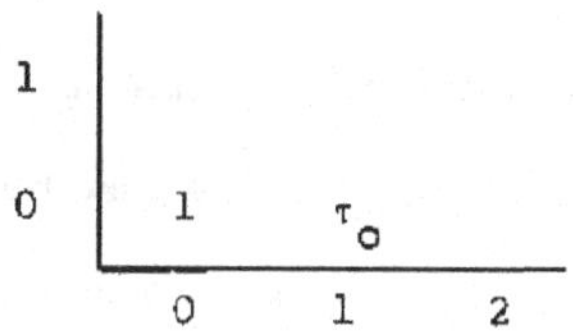

Figure 1

Thus τ_o is an infinite cycle, as is the identity, 1.

For convenience let us write $E^r(\underline{K};\underline{L}_p)$ as E^r with its differential d^r, and $E^r(\underline{K}_p;\underline{L}_p)$ as ${}^pE^r$ with its differential d^r_p. We know that ${}^pE^2 = E^2 \oplus \tau_o E^2$. Assume that ${}^pE^r = E^r \oplus \tau_o E^r$, $r \geq 2$. Then first we note that by the derivation property (Theorem 5.9) and by the fact that $d^r_p(\tau_o) = 0$ for all $r \geq 2$, we have that $d^r_p(\tau_o a) = \pm\tau_o d^r_p(a)$. Let $d^r_p(a) = b + \tau_o c$, with $a,b,c \in E^r$. Then take β of both sides. This yields $0 = \pm d^r_p(\beta a) = \beta b + \beta(\tau_o c) = c$ since $\beta(E^r) = 0$. Thus $d^r_p(a) = b$ so $d^r_p(E^r) \subset E^r$ and by naturality we see that $d^r_p|E^r = d^r$. Thus ${}^pE^{r+1} = E^{r+1} \oplus \tau_o E^{r+1}$. Inductively, then, the whole spectral sequence splits into two isomorphic pieces, one of which is $\{E^r, d^r\}$ the other, τ_o times the first. So we may as well consider only the spectral sequence E^r.

Let us look at it for $p = 3$:

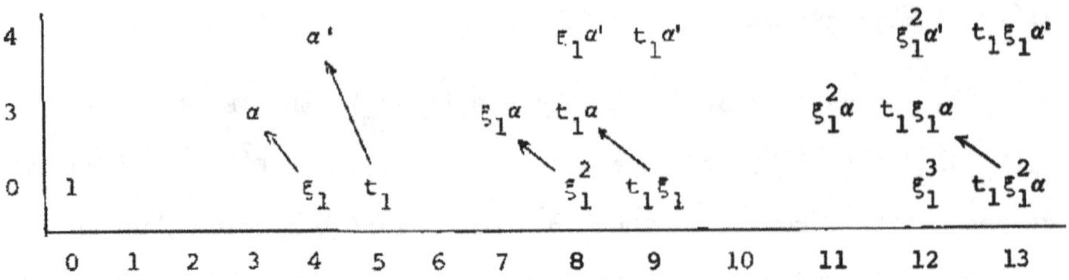

Figure 2

First we note that there is some $\alpha \in \pi_3$ so that ξ_1 can cancel, $\alpha' \in \pi_4$ to cancel t_1. (We saw before that α comes from something of order 3 in $\pi_3(\underline{S})$. Thus $\partial_3(\alpha') = \alpha$.)

Next we see that $d^4(\xi_1^2) = 2\xi_1 d^4(\xi_1) = 2\xi_1\alpha$, $d^4(t_1\xi_1) = t_1 d^4(\xi_1) = t_1\alpha$, $d^4(t_1\xi_1^2) = 2t_1\xi_1 d^4(\xi_1) = 2t_1\xi_1\alpha$. Thus cancellation takes place as indicated by the arrows. Rather than doing cancellation one term at a time, let us recall that d^4 going from the base to line 3 (which we indicated by $d^4: 1 \to \alpha$) must represent a primary homology operation. In this case since $d^4(\xi_1) = \alpha$, $d^4 = \rho^1$. We can use Theorem 5.8 to calculate the action of ρ^1 on A_* and find

$$\ker \rho^1 = P(\xi_1^3, \xi_2, \ldots) \otimes E(t_1, t_2, \ldots) \quad \text{which we call } D$$

$$\operatorname{cok} \rho^1 = \xi_1^2 D .$$

Observe that in general $d^2 = d^3 = 0$ so that d^4 is defined everywhere. If $\theta \in \pi_*(\underline{L}_3)$ and $\theta\alpha \neq 0$ then $d^4: \theta \to \theta\alpha$ is ρ^1. Thus after d^4 acts we have left $D\theta$ and $\xi_1^2 D\theta\alpha$ (the rest having cancelled).

Next we look at $d^5: 1 \to \alpha'$. d^5 must be represented by some primary operation $T \in \mathcal{G}^5$. $T = \lambda P^1\beta + \mu\beta P^1$, $\lambda,\mu \in Z_3$ where $t_1 T = 1$. Since T will only be operating on $\ker P^1$ we may as well assume $\lambda = 0$ as its value does not matter. Now $t_1\beta P^1 = -1$ so we have $\mu = -1$. $\xi_n T = 0$, $t_{n+1}T = 0$ for all $n \geq 1$. Thus if we write $\ker P^1 = D = D' \oplus t_1 D'$ where $D' = P(\xi_1^3, \xi_2, \ldots) \otimes E(t_2, t_3, \ldots)$, then the action of T here is to map D' to 0 and $t_1 D'$ isomorphically onto D'.

Look back at Figure 2. Notice that $\xi_1\alpha'$ and $t_1\alpha'$ are unbounded. Thus they must have non-zero boundaries: $0 \neq d^4(\xi_1\alpha')$, $0 \neq d^5(t_1\alpha')$. From the derivation property then, $\alpha'\alpha \neq 0$, $(\alpha')^2 \neq 0$.

But then from what we know about $d^4: 1 \to \alpha$ and $d^4: \alpha' \to \alpha'\alpha$, $d^5: 1 \to \alpha'$ and $d^5: \alpha' \to (\alpha')^2$ we see that we are left with D' in line 0; $\xi_1^2 D\alpha$ in line 3; 0 in line 4. This last is the most interesting. At this point, observe that "it would be nice" if $(\alpha')^n$ and $(\alpha')^n\alpha$ were non-zero. For then we could get much cancellation in a trivial manner (all the $(\alpha')^n$ and $(\alpha')^n\alpha$ would be cancelled for $n \geq 1$). In fact this turns out to be true. The proof is too deep for us to go into here, but the idea is as follows:

The map $\alpha': S^{n+4} \to S^n \cup_3 e^{n+1} = M(Z_3, n)$ is of order 3, hence it can be extended to a map $\hat{\alpha}: S^{n+4} \cup_3 e^{n+5} \to S^n \cup_3 e^{n+1}$. Let C be the mapping cone of $\hat{\alpha}$. Then J. F. Adams shows (in <u>Adams</u> [3]) that $K(C) = K(*)$ where K is complex K-theory, the cohomology theory whose spectrum is U, $BU \times Z$, $U, \ldots$. Thus $K(\hat{\alpha})$ is an isomorphism (and is non-trivial). Thus considering $\hat{\alpha} \in [\underset{\sim}{L}_3, \underset{\sim}{L}_3]_4$ we must have $K(\hat{\alpha}^n)$ an isomorphism for all n so $\hat{\alpha}^n \neq 0$. It follows that $(\alpha')^n\alpha$ and $(\alpha')^n$ are non-trivial as elements of $\pi_*(\underset{\sim}{L}_3)$. (This goes through exactly the same way for all primes

$p > 3$ for $\alpha \in \pi_{2p-3}(\underline{L}_p)$.)

We then get the following over-all cancellation:

$$(\xi_1 D \oplus \xi_1^2 D)(\alpha')^n \xrightarrow{d^4} (D \oplus \xi_1 D)(\alpha')^n \alpha ,$$

$n \geq 0$ leaving $D(\alpha')^n$ and $\xi_1^2 D(\alpha')^n \alpha$, $n \geq 0$. Next $t_1 D'(\alpha')^n \xrightarrow{d^5} D'(\alpha')^{n+1}$, $t_1 D'(\alpha')^n \alpha \xrightarrow{d^5} D'(\alpha')^{n+1} \alpha$ leaving only D' in degree 0 and $\xi_1^2 D'\alpha$ in degree 3 .

The next non-zero elements remaining are as in Figure 3:

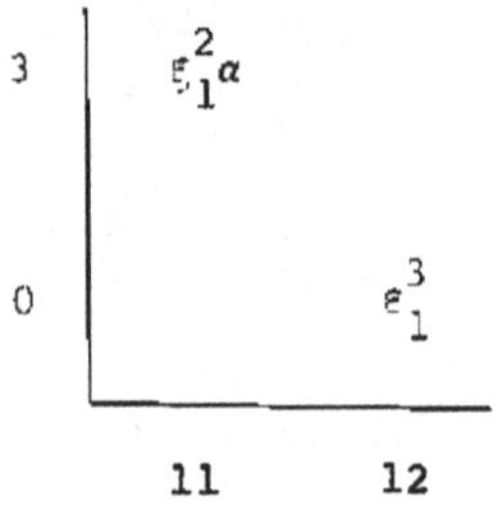

Figure 3

In order to cancel, then, we must have $d^8(\xi_1^2 \alpha) \neq 0$, $d^{12}(\xi_1^3) \neq 0$. Let these new elements be $\beta_1 \in \pi_{10}(\underline{L}_3)$, $\beta_1' \in \pi_{11}(\underline{L}_3)$. One really needs to check that $\beta_1 \in \pi_{10}(\underline{S})$ is of order 3. This comes from checking back with the spectral sequence $E^r(\underline{K};\underline{S})$. This justifies calling the element β_1' with $\partial_3(\beta_1') = \beta_1$.

We next note that $d^8 : \alpha \to \beta_1$ must be ρ^2 . Thus we get $\xi_1^2 D'\alpha \xrightarrow{d^8} D'\beta_1$. Further, $d^{12} : 1 \to \beta_1'$ must be ρ^3 .

The next non-zero elements are

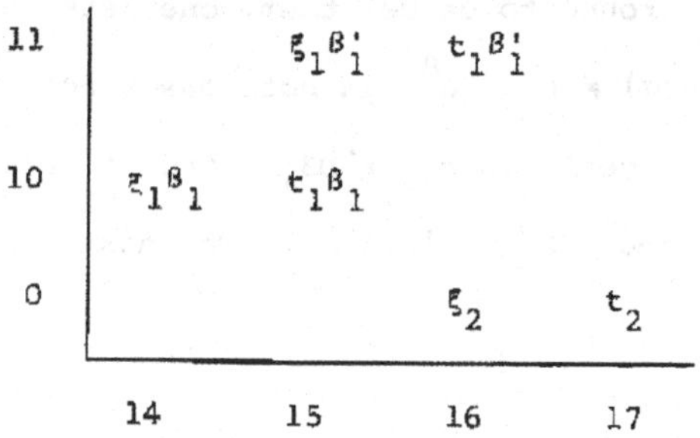

Figure 4

Now $d^{12}(t_2) = (t_2\rho^3)\beta_1' = t_1\beta_1'$. Thus $0 = d^5(t_1\beta_1') = \alpha'\beta_1'$. Applying ∂_3 to this equation yields $0 = \partial_3(\alpha')\beta_1' + \alpha'\partial_3(\beta_1') = \alpha\beta_1' + \alpha'\beta_1$. So $\alpha'\beta_1 = -\alpha\beta_1'$. Thus $d^5(t_1\beta_1) = -d^4(\xi_1\beta_1')$, hence is zero in E^5 . Thus $t_1\beta_1$ is an infinite cycle, thus $d^{11}(\xi_2) = t_1\beta_1$ (up to non-zero multiple). Then since $\xi_1\beta_1$ and $\xi_1\beta_1'$ are unbounded, $0 \neq d^4(\xi_1\beta_1) = \alpha\beta_1$, $0 \neq d^4(\xi_1\beta_1') = \alpha\beta_1'$.

Then we have cancellation occurring:

$$(\xi_1 D \oplus \xi_1^2 D)\beta_1 \xrightarrow{d^4} (D \oplus \xi_1 D)\beta_1\alpha$$

$$(\xi_1 D \oplus \xi_1^2 D)\beta_1' \xrightarrow{d^4} (D \oplus \xi_1 D)\beta_1'\alpha$$

leaving altogether

$$D'; \quad t_1 D'\beta_1; \quad D\beta_1'; \quad \xi_1^2 D\beta_1\alpha; \quad \xi_1^2 D\beta_1'\alpha .$$

(We do not do a general cancellation for d^{11} and d^{12} as this is very complicated).

The next terms will be $\xi_1^2\beta_1\alpha$ and $\xi_1^2\beta_1'\alpha$. Since there is nothing else around to cancel them, one gets $\beta_1^2 = d^8(\xi_1^2\beta_1\alpha) \neq 0$,

$\beta_1'\beta_1 = d^8(\xi_1^2\beta_1'\alpha) \neq 0$. d^8 in both cases acts like ρ^2 .

This last gets rid of $\xi_1^2 D\beta_1\alpha$ and $\xi_1^2 D\beta'\alpha$ but adds

$(A_* - D)\beta_1^2$, and $(A_* - D)\beta_1'\beta_1\alpha$. The next non-zero terms to appear

are these:

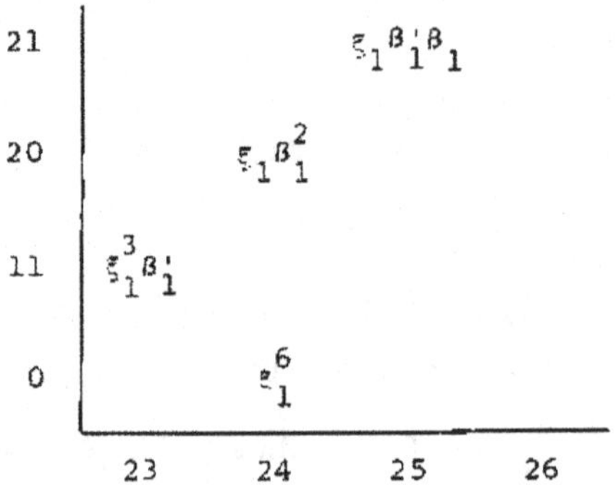

Figure 5

Clearly $d^{12}(\xi_1^6) = d^{12}((\xi_1^3)^2) = 2\xi_1^3 \cdot d^{12}(\xi_1^3) = 2\xi_1^3\beta_1'$. Then we get in

addition $\beta_1^2\alpha$ and $\beta_1'\beta_1\alpha$ non-zero. So $(A_* - D)\beta_1^2$ and

$(A_* - D)\beta_1'\beta_1\alpha$ cancel and we have the new terms $\xi_1^2 D\beta_1^2\alpha$ and

$\xi_1^2 D\beta_1'\beta_1\alpha$. The next terms to arise are these:

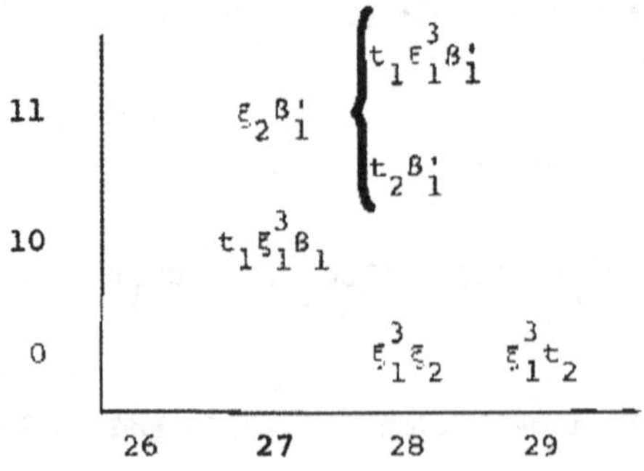

Figure 6

We have $d^{11}(\xi_1^3\xi_2) = \xi_1^3 d^{11}(\xi_2) = t_1\xi_1^3\beta_1$ and $d^{12}(\xi_1^3 t_2) = d^{12}(\xi_1^3)t_2 + \xi_1^3 d^{12}(t_2) = -t_2\beta_1' + t_1\xi_1^3\beta_1'$. This leaves $\xi_2\beta_1'$ and $t_2\beta_1'$ generating the unbounded terms. Thus $\beta_2 = d^{16}(\xi_2\beta_1') \neq 0$ and $\beta_2' = d^{17}(t_2\beta_1') \neq 0$. (Again, one must check $E^r(\underline{K};\underline{S})$ that $\partial_3(\beta_2') = \beta_2$.)

The next picture is this:

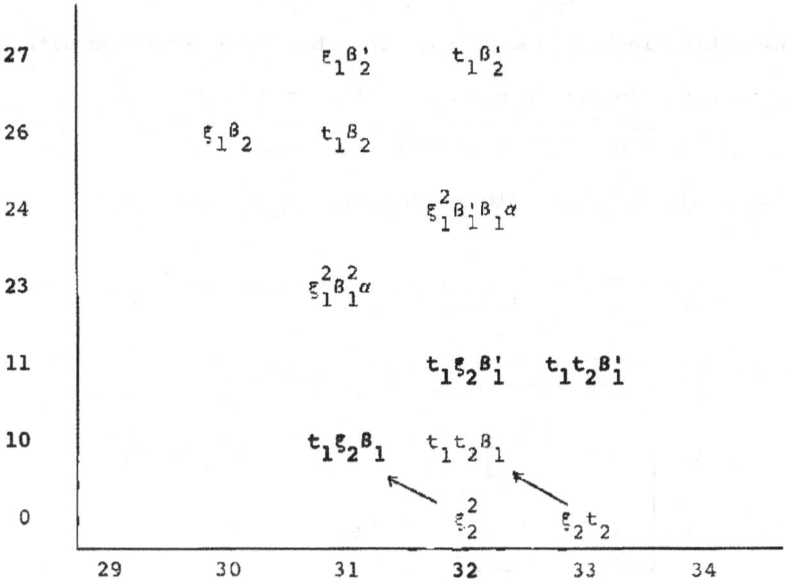

Figure 7

We get the cancellation indicated by the arrows easily. We get $d^4(\xi_1^2\beta_1^2\alpha) = -\xi_1\beta_1^2\alpha\cdot\alpha = 0$ so that $0 \neq d^4(\xi_1\beta_2) = \beta_2\alpha$. Thus also $d^4(\xi_1\beta_2') = -\beta_2'\alpha \neq 0$ (since $\partial_3(\beta_2'\alpha) = \beta_2\alpha$). We next note that we have a choice of cancellations: if $d^{13}(t_1\xi_1\beta_1') = \pm\,\xi_1^2\beta_1^2\alpha$ then $\beta_2\alpha' \neq 0$; if $d^{13}(t_1\xi_2\beta_1') = 0$ then $d^{16}(t_1\xi_2\beta_1') =$ some multiple of $t_1\beta_2'$.

Using naturality of left acting homology operations we have

$\beta P^1 d^{16}(t_1\xi_2\beta_1') = d^{16}(\beta P^1 t_1\xi_2\beta_1')$; since $\beta P^1 t_1\xi_2 = \xi_2 + \xi_1^4$ and $\beta P^1 t_1 = 1$ this gives $d^{16}(t_1\xi_2\beta_1') = t_1\beta_2$. Similarly $t_1t_2\beta_1'$ cancels either $t_1\beta_2'$ or $\xi_1^2\beta_1'\beta_1\alpha$.

Thus we have either $\beta_2\alpha'$ independent of $\beta_2'\alpha$ and $\beta_1^3 = 0$ modulo $(\beta_2'\alpha, \beta_2\alpha')$ or $\beta_2\alpha'$ dependent on $\beta_2'\alpha$ and β_1^3 independent of $\beta_2'\alpha$.

So we know that $\pi_{30}(\underline{L}_3) \cong Z_3 \oplus Z_3$ and $\pi_{31}(\underline{L}_3) \cong Z_3$ but we do not know where the elements come from. Toda has some recent results which imply that the former is the case: $\beta_2'\alpha$ and $\beta_2\alpha'$ are independent and $\beta_1^3 \neq 0$ is a combination of them.

The next stage of the problem brings us to

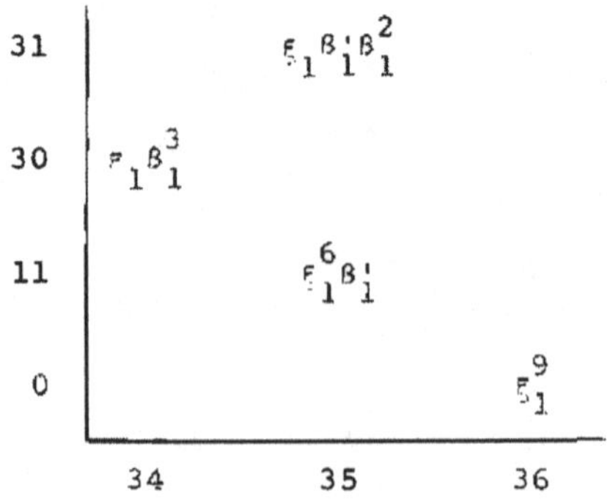

Figure 8

Again we do not know how cancellation occurs. The solution here is much deeper. It turns out - again due to Toda - that $d^{20}(\xi_1^6\beta_1') = \pm\, \xi_1\beta_1^3$, $d^{32}(\xi_1^9) = \pm\, \xi_1\beta_1'\beta_1^2$ so that $\beta_1^3\alpha = 0$, $\beta_1'\beta_1^2\alpha = 0$.

The history of these last two differentials is the following: Toda [1] had given a false proof that $\beta_2\alpha' = \beta_2'\alpha$. Based on that relation and using these methods, one can compute the differentials of the above diagram (except that the wrong conclusion is reached). This author used these methods, then, to "compute" the differential of Figure 8, showing that $d^{20}(\xi_1^6\beta_1') = 0$ and $d^{32}(\xi_1^9) = 0$ whence $\beta_1^3\alpha \neq 0$. Toda then came across a contradiction when his independent methods showed $\beta_1^3\alpha = 0$. Working backwards it was discovered that the error in Cohen [1] followed from the error in Toda [1].

Everything here generalizes immediately to all primes $p > 3$. In general $\pi_*(\underline{L}_p)$ has ring generators $\alpha \in \pi_{2p-3}(\underline{L}_p)$, $\alpha' \in \pi_{2p-2}(\underline{L}_p)$, $\beta_r \in \pi_{2(p-1)(rp+r-1)-2}(\underline{L}_p)$, $\beta_r' \in \pi_{2(p-1)(rp+r-1)-1}(\underline{L}_p)$, $r = 1,\ldots,p-1$ graded commutative with relations $\beta_r\alpha' = \beta_r'\alpha$, $r = 1,\ldots,p-1$; (for $r = p-1$ this differs from the case $p = 3$), $\beta_1^p\alpha = 0$, $\beta_1'\beta_1^{p-1}\alpha = 0$. (Other relations follow: e.g. $\beta_r\alpha'\alpha = \beta_r'\alpha^2 = 0$ for $1 \leq r \leq p-1$.)

As in the case of $p = 3$, $\beta_1 = d^{2(p-1)^2}(\xi_1^{p-1}\alpha)$, $\beta_{r+1} = d^{2(p^2-1)}(\xi_2\beta_r')$, $r \geq 1$.

This can all be put into $\pi_*(\underline{S})$ and summed up as follows:

<u>Theorem 5.11</u>: The p-primary component of $\pi_n(\underline{S})$ for $0 < n \leq (p^2 + 2p)q - 6$ (where p is an odd prime and $q = 2(p-1)$) is described completely by listing the following basis elements: first there are elements α_n in degree $nq - 1$ of order p but divisible by n and $\frac{1}{n}\alpha_n$ generates a summand. There is an element φ of degree $(p^2 + p)q - 3$ which is of order p^2. The following generate summands of order p: $\beta_1^r\beta_s$, $\beta_1^r\beta_s\alpha_1$, $\beta_1^r\epsilon_n'$, ϵ_m, $\beta_2\beta_{p-1}$, $r \geq 0$; $s,m = 1,\ldots,p-1$; $n = 1,\ldots,p-2$. The degrees are given by $\deg \beta_s = (sp + s - 1)q - 2$, $1 + \deg \epsilon_m' = \text{degree } \epsilon_m = (p^2 + m)q - 2$ and $\beta_1^p\alpha_1 = 0$. There is a relation $\epsilon_{p-1}\alpha_1 = p\varphi$.

The reader will observe that these results go to a much higher degree than the calculations. This is so that anyone wishing to try these methods may have a check on his results. The methods outlined (feeding in $\beta_1^p \alpha_1 = 0$) will allow one to calculate up to the indicated degree.

5.5 Other Results

Beside the ordinary composition product in $\pi_*(\underline{S})$, there are some "higher products" that can be defined.

Assume that we have a sequence of maps of spaces

$$A \xrightarrow{f} B \xrightarrow{g} C \xrightarrow{h} D$$

in which $hgf \sim *$ for two different reasons: namely $gf \sim *$ and $hg \sim *$. We wish to get a new map reflecting both facts.

Recall that $f: X \to Y$ is null homotopic if and only if there is a map $\varphi: TX \to Y$ with $\varphi|X = f$. Now in our case if $gf: A \to C$ is $\sim *$ then there is some $\alpha: TA \to C$ with $\alpha|A = gf$. Thus for $h\alpha: TA \to D$ we have $h\alpha|A = hgf$. Furthermore $hg \sim *$ so there exists $\beta: TB \to D$ with $\beta|B = hg$. Then for $\beta\circ(Tf): TA \to D$ we have $\beta\circ(Tf)|A = hgf$.

But now let us consider SA as two cones over A fitted together at the base: i.e. $T^{+}A = \mathrm{im}([\frac{1}{2},1] \times A \to SA)$, $T^{-1}A = \mathrm{im}([0,\frac{1}{2}] \times A \to SA)$, $SA = T^{+}A \cup T^{-1}A$, $T^{+}A \cap T^{-1}A = \frac{1}{2} \times A$. Then define $\varphi: SA \to D$ by $\varphi|T^{+}A = h\alpha$, $\varphi|T^{-}A = \beta(Tf)$ (they both agree on the intersection). The homotopy classes of all such maps is $\langle h,g,f \rangle \subset [SA,D]$. This is called the triple Toda bracket.

Algebraically we have the following diagram

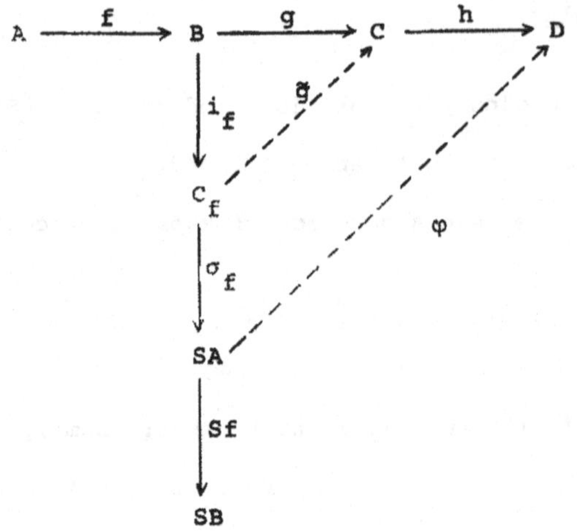

Since $gf \sim *$, $\tilde{g}$ exists with $\tilde{g}i_f \sim g$. Then $(h\tilde{g})i_f \sim hg \sim *$, so there exists φ with $\varphi\sigma_f \sim h\tilde{g}$.

The set of all such $[\varphi]$ is $\langle h,g,f\rangle$. First we observe that $[\tilde{g}]$ is chosen with indeterminacy $\sigma_f^*[SA,C]$; i.e. $[\tilde{g}_o] - [\tilde{g}_1] = \sigma_f^*\alpha$ for some $\alpha \in [SA,C]$ if $\tilde{g}_o$ and $\tilde{g}_1$ are two choices for $\tilde{g}$. For fixed $\tilde{g}$, $[\varphi]$ is chosen with indeterminacy $(Sf)^*[SB,D]$. Thus $[\varphi_o] - [\varphi_1] = (Sf)^*\beta$ for some $\beta \in [SB,D]$ if φ_o and φ_1 are choices for φ such that $\varphi_o\sigma_f \sim \varphi_1\sigma_1 \sim h\tilde{g}_1$. But then choosing a different $\tilde{g}$, say $[\tilde{g}_o] = [\tilde{g}_1] + \sigma_f^*\alpha$ clearly $[\varphi_o] + \alpha$ will do, so the total indeterminacy in the choice of φ is $h_*[SA,C] + (Sf)^*[SB,D]$. (And if φ_o is a choice and φ_1 differs from φ_o by an element of this subgroup then φ_1 will be in $\langle h,g,f\rangle$). Thus $\langle h,g,f\rangle$ is a coset of $h_*[SA,C] + (Sf)^*[SB,D]$ in $[SA,D]$.

In the stable range we have an alternate definition:

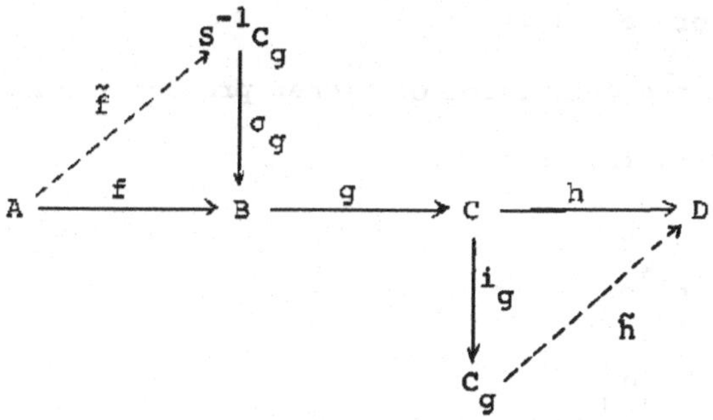

Since $gf \sim *$ and $hg \sim *$ we can define $\tilde{f}$ and $\tilde{h}$ making the diagram commute. Then $\tilde{h}\circ(S\tilde{f}):SA \to D$ is defined and we observe that it has exactly the right indeterminacy. By "stable range," we mean at least that g is a suspension and that

$[A,S^{-1}C_g] \to [A,B] \to [A,C]$ is exact.

For example in $\pi_*(\underline{S})$:

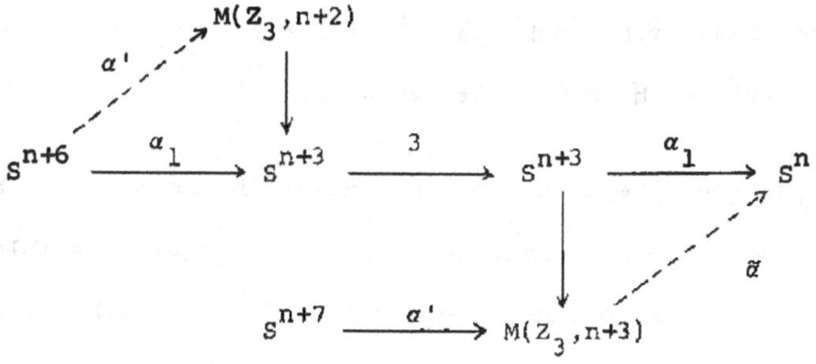

$\tilde{\alpha}\circ\alpha':S^{n+7} \to S^n$ yields $0 \neq \alpha_2 \in \pi_7(\underset{\sim}{S})$. $\alpha_2 = \langle\alpha_1,3,\alpha_1\rangle$ (no indeterminacy), is of order 3 and its image in $\pi_*(\underline{L}_3)$ is $\alpha'\alpha$. In general $\langle\alpha_n,p,\alpha_1\rangle$ is defined and yields a non-zero element α_{n+1} of order p . It turns out that α_n is divisible by n , $\tilde{\alpha}_n = \frac{1}{n}\alpha_n$. Then the image of $\tilde{\alpha}_n$ in $\pi_*(\underline{L}_p)$ is $(\alpha')^n\alpha$ and $\partial((\alpha')^n) = \alpha_n$, $\partial:\pi_*(\underline{L}_p) \to \pi_*(\underline{S})$.

In addition, among the elements we have met it turns out that

$\beta_1 = \langle\alpha_1,\alpha_1,\alpha_1\rangle$, for $p = 3$.

One can extend the definition of higher products to an n-fold bracket. For example, for $n = 4$:

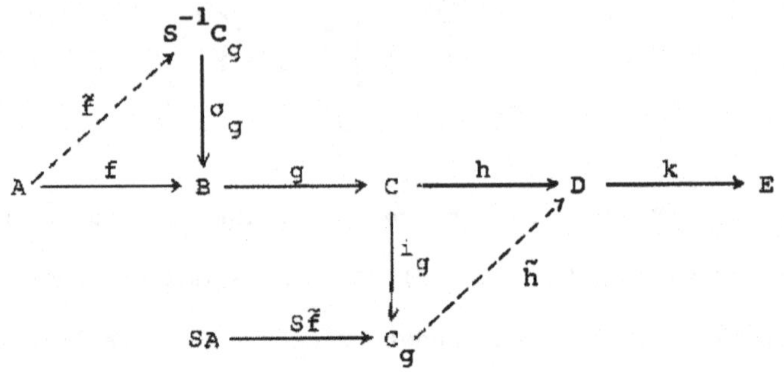

If, first of all, all composites of two are null homotopic, and both triple products <u>contain</u> 0 , then we know that there exist liftings $\tilde{f},\tilde{h}$ and

a) there exist $\tilde{f},\tilde{h}$ such that $\tilde{h}\circ S\tilde{f} \sim *$;

b) there exists $\tilde{h}$ such that $k\tilde{h} \sim *$.

If, in addition, there is some $\tilde{h}$ satisfying both a) and b), then $\langle k,h,S\tilde{f}\rangle \subset [S^2A,D]$ is defined. The union of all possible such sets is $\langle k,h,g,f\rangle$. Important: $\langle k,h,g,f\rangle \neq \emptyset \Rightarrow 0 \in \langle k,h,g\rangle$ and $0 \in \langle h,g,f\rangle$ but the converse does not hold. So higher products are very messy, but can be defined.

In particular we have a few useful results based on these higher products.

<u>Theorem 5.12</u>: Everything in $\pi_*(\underline{S})$ is decomposable into higher Toda brackets (and Toda brackets of Toda brackets, etc.) in terms of $\alpha_1 \in \pi_{2p-3}(\underline{S})$ for odd primes p and the 2-primary elements

$\eta \in \pi_1(\underline{S})$, $\nu \in \pi_3(\underline{S})$ and $\sigma \in \pi_7(\underline{S})$.

The idea behind the proof (details of which are in Cohen [2]) is to look at $E^r(\underline{K};\underline{S})$ and to prove that if $1 \otimes \theta \in E^2_{o,n} \cong \pi_n(\underline{S})$ is an r-boundary of something for $r \leq n$ then it is a Toda bracket of lower dimensional elements. Then it is proved that the above-mentioned elements are the only ones which are in the image of $d^{n+1}:E^{n+1}_{n+1,0} \to E^{n+1}_{o,n}$. We have a similar sort of decomposition in the spectral sequence $E^r(\underline{K};\underline{X})$, $\underline{X}$ any spectrum, leading to the fact that all of $\pi_*(\underline{X}) \cong E^2_{o,*}$ is generated by Toda brackets of elements of $\pi_*(\underline{S})$ acting on $E^\infty_{o,*}$, a certain quotient group of $\pi_*(\underline{X})$ which has the following geometric description:

Let $C_*(\underline{X}) = \mathrm{im}(h:\pi_*(\underline{X}) \to H_*(\underline{X}))$ where h is the Hurewicz homomorphism. If $\iota \in [\underline{S},\underline{K}]_o$ is the generator then

$$\begin{array}{ccc} \iota_*:\pi_*(\underline{S} \wedge -) & \to & \pi_*(\underline{K} \wedge -) \\ \| & & \| \\ \pi_*(-) & \to & H_*(-) \end{array}$$

is h . Thus $\iota \in E^\infty_{o,o}(\underline{K};\underline{S})$ generates all of E^∞_{**} where ι is the image of $1 \otimes 1 \in E^2_{o,o}(\underline{K};\underline{S})$. Given any $\theta \in \pi_*(\underline{X})$, $\theta:\underline{S} \to \underline{X}$, we have $h(\theta) = \iota_*(\theta) = \theta_*(\iota)$ so since $\theta_*[1 \otimes 1] = 1 \otimes \theta$ in $E^2_{o,*}$, clearly the image of $1 \otimes \theta$ in $E^\infty_{o,*} \subset H_*(\underline{X})$, is $h(\theta)$. Thus $E^\infty_{o,*} = C_*(\underline{X})$ and we get

<u>Theorem 5.13</u>: From the Toda bracket construction on $C_*(\underline{X})$ we can construct all of $\pi_*(\underline{X})$.

We can also use C_* to get another interesting relation. First we mention the following lemma:

<u>Lemma 5.14</u>: Let $\{E^r_{**}, d^r\}$ be a first quadrant Serre-type spectral sequence (i.e. $E^2_{s,t} \simeq E^2_{s,o} \otimes E^2_{o,t} \oplus \mathrm{Tor}(E^2_{s-1,o}, E^2_{o,t})$) with each $E^2_{s,t}$ finitely generated. Let $f: E^r_{**} \to E^r_{**}$ be an endomorphism such that

1) $f|E^2_{*,o}$ is an automorphism

2) $f|E^\infty_{o,*}$ is an automorphism

then f is an automorphism.

The proof is purely an exercise in algebraic manipulation. For the details see <u>Cohen</u> [3].

Let $f:\underline{X} \to \underline{X}$ be a self-map of a convergent spectrum of finite type. Assume $f_*: C_*(\underline{X}) \to C_*(\underline{X})$ is an isomorphism. Then $f_*: E^r(\underline{K};\underline{X}) \to E^r(\underline{K};\underline{X})$ induces the identity in the bottom row (which by a shift in dimension we may take to be $E^2_{*,o}$ and $f_*: E^\infty_{o,*} \to E^\infty_{o,*}$ is an isomorphism since $E^\infty_{o,*} \simeq C_*(\underline{X})$ hence by the lemma $f_*: \pi_*(\underline{X}) \to \pi_*(\underline{X})$ is an isomorphism so f is a weak homotopy equivalence.

<u>Corollary 5.15</u>: Let $\underline{X}$ and $\underline{Y}$ be convergent spectra of finite type (homology finitely generated in each degree). If $f:\underline{X} \to \underline{Y}$ and $g:\underline{Y} \to \underline{X}$ induce C_*-isomorphisms they are weak homotopy equivalences.

For X and Y CW complexes of finite type we can define $X \leq Y$ if there exists $f: X \to Y$ with $C_*(f)$ an isomorphism and $X = Y$ if there exists $f: X \to Y$ with $\pi^s_*(f)$ an isomorphism. Then $\leq$ is a partial ordering by this corollary.

5.6 The Adams Spectral Sequence

We now discuss the Adams Spectral Sequence. Fix a prime p throughout this section. Let $H^*(\ ) = H^*(\ ;Z_p)$.

Definition: A free spectrum $\underline{F}$ is a sum of various non-negative suspensions of $\underline{K}_p$, only finitely many in each degree. I.e.,
$\underline{F} = \bigvee_{i=0}^{\infty} \bigvee_{j=1}^{n_i} \underline{K}_p^i$.

Observe that if $\underline{F}$ is free then $H_*(\underline{F})$ is a free G^*-module of finite type with generators in non-negative degrees. (From now on we will just say free G^*-module with the rest to be understood.) Conversely, if C is any free G^*-module then there is a free spectrum $\underline{F}_C$ with $H^*(\underline{F}_C) \cong C$ as an G^*-module. In fact $C \mapsto \underline{F}_C$ is a contravariant functor since clearly an G^*-morphism $C \to D$ yields a map $\underline{F}_D \to \underline{F}_C$. Furthermore, for any spectrum $\underline{W}$, $[\underline{W},\underline{F}_C] \cong \mathrm{Hom}_{G^*}(C,H^*(\underline{W}))$, a natural isomorphism.

Let $\underline{X}$ be a convergent spectrum of finite type - for simplicity assume $\underline{X}$ is (-1)-connected. Then $H^*(\underline{X})$ is an G^*-module of finite type with generators in non-negative degrees so we can get a free resolution.

$$0 \longleftarrow H^*(\underline{X}) \longleftarrow C_o \xleftarrow{g_o} C_1 \xleftarrow{g_1} C_2 \longleftarrow \cdots$$

Applying F and using the previous, this yields a sequence of maps

$$0 \to \underline{X} \xrightarrow{f_{-1}} \underline{F}_o \xrightarrow{f_o} \underline{F}_1 \xrightarrow{f_1} F_2 \to \cdots$$

where the $\underline{F}_i$ are free.

We shall define a sequence of spectra $\underline{M}^t$ and exact triangles (i.e. cofibrations) of spectra

$$\begin{array}{ccc} & \underline{F}^t & \\ {\scriptstyle \alpha_t}\nearrow & & \searrow{\scriptstyle \beta_t} \\ \underline{M}^{t-1} & \xleftarrow{\ \gamma\ } & \underline{M}^t \end{array}$$

such that $\underline{M}^t = \underline{C}_{\alpha_t}$ and $H^*(\beta_{t-1})$ is a monomorphism (equivalently $H^*(\alpha_{t-1})$ is an epimorphism) and $\alpha_{t+1}\beta_t \sim f_t$.

Let $\underline{M}^{-1} = \underline{X}$, $\alpha_o = f_{-1}$, $H^*(\alpha_o) = \epsilon$ an epimorphism. $\alpha_{-1}:0 \to \underline{X}$ clearly has $H^*(\alpha_{-1})$ an epimorphism.

Assume $\underline{M}^s$ is defined and $H^*(\beta_s) = \beta^*_s$ is a monomorphism for $s \leq t$. Then $\beta^*_{t-1}:[\underline{M}^{t-1},\underline{F}^{t+1}] \to [\underline{F}^{t-1},\underline{F}^{t+1}]$ is a monomorphism, so since $\beta^*_{t-1}[f_t\alpha_t] = [f_tf_{t-1}] = 0$ we have $f_t\alpha_t \sim 0$ thus f_t may be factored $f_t \sim \alpha_{t+1}\beta_t$. We wish to show that $H^*(\alpha_{t+1})$ is an epimorphism: since $\beta_tf_{t-1} \sim (\beta_t\alpha_t)\beta_{t-1} \sim *$ we have $\beta^*_tH^*(\underline{M}^t) \subset \ker f^*_{t-1} = \operatorname{im} f^*_t = \beta^*_t \operatorname{im} \alpha^*_{t+1}$. But β^*_t is a monomorphism hence $H^*(\underline{M}^t) = \operatorname{im} \alpha^*_{t+1}$ so α^*_{t+1} is an epimorphism and the construction is complete.

Thus for any spectrum $\underline{Y}$ we get a bigraded exact triangle by summing up the following for all t :

$$\begin{array}{ccc} [\underline{Y},\underline{M}^t]^* & \xrightarrow{\ \gamma_{t*}\ } & [\underline{Y},\underline{M}^{t-1}]^* \\ {\scriptstyle \beta_{t*}}\nwarrow & & \swarrow{\scriptstyle \alpha_{t*}} \\ & [\underline{Y},\underline{F}^t]^* & \end{array}$$

in which $E_1^{t*} = [\underline{Y},\underline{F}^t]^*$ and $d_1^{t,*}:E_1^{t*} \to E_1^{t+1,*}$ is $\alpha_{t+1*}\beta_{t*} = f_{t*}$.

Thus $E_1^{t*} \simeq \mathrm{Hom}_{G*}(C_t, H^*(\underline{Y}))$ and d_1^{t*} is g_t^*. Thus by definition

$$E_2^{**} = \mathrm{Ext}_{G*}^{**}(H^*(\underline{X}), H^*(\underline{Y}))$$

and as usual if we let

$$G_{n*} = \mathrm{im}([\underline{Y},\underline{M}^n]* \xrightarrow{\gamma_{n*}} [\underline{Y},\underline{M}^{n-1}]* \xrightarrow{\gamma_{n-1}} \cdots \to [\underline{Y},\underline{X}]*)$$

we get that E_∞^{n*} is G_{n*}/G_{n+1*} so that E^{**} gives the quotients in a filtration of

$$\frac{[\underline{Y},\underline{X}]}{\bigcap_n G_n} .$$

If $\underline{Y}$ is a finite spectrum, then it can be shown that $\cap G_n$ contains just what it should: Observe that each $\underline{F}^t$ has its identity map of order p. For any group π and prime q let ${}^q\pi = \{\theta \in \pi | \theta$ is of order q^n for some $n\}$. Let ${}_p\pi = \pi / \sum_{q\neq p} {}^q\pi$. Then $p \cdot [\underline{Y},\underline{F}^t]* = 0$. Thus for any prime $q \neq p$ ${}^q[\underline{Y},\underline{M}^t]* \xrightarrow{\gamma_{t*}} {}^q[\underline{Y},\underline{M}^{t-1}]*$ is an isomorphism. Thus $\bigcap_n G_n \supset \sum_{q\neq p} {}^q[\underline{Y},\underline{SX}]$.

It can be proved that for $\underline{Y}$ the S-spectrum of a finite CW complex we get equality. Thus E_∞^{**} yields the quotients in a filtration of ${}_p[\underline{Y},\underline{X}]$. Thus we get the Adams Spectral Sequence. (For details see Mosher-Tangora or Adams [1] or [2].)

Theorem 5.16: Let $\underline{Y}$ be a finite spectrum and $\underline{X}$ be a convergent spectrum of finite type and G^* be the mod p Steenrod Algebra. Then there is a spectral sequence converging to quotients in a filtration of ${}_p[\underline{Y},\underline{X}]$ in which $E_2^{**} \simeq \mathrm{Ext}_{G*}^{**}(H^*(\underline{X};Z_p), H^*(\underline{Y};Z_p))$.

5.7 Graded Localization

Throughout this section R will be a commutative graded ring with unit $1 \in R_0$. Thus, if $x \in R_n$ and $y \in R_m$, $xy = (-1)^{nm} yx$. If M and N are graded R-modules, an R-homomorphism $f: M \to N$ of degree d is a sequence of homomorphisms $f_n: M_n \to N_{n+d}$ such that for $r \in R_t$, and $m \in M_n$, $f_{n+t}(rm) = (-1)^{td} r\, f_n(m)$. Let T be a multiplicatively closed subset of R. Multiplicatively closed will always imply that $1 \in T$ and $0 \notin T$.

If M is an R-module, we define M_T to be the set of formal quotients $\frac{m}{t}$ for $m \in M$, $t \in T$ with the relation $\frac{m}{t} = \frac{m'}{t'}$ if for some $t'' \in T$, $t''(t'm - tm') = 0$. Deg $\frac{m}{t}$ = deg m - deg t, of course.

Define addition by $\frac{m}{t} + \frac{m'}{t'} = \frac{t'm + tm'}{tt'}$. In particular R_T becomes a ring under the definition $\frac{r}{t} \cdot \frac{r'}{t'} = \frac{rr'}{tt'}$. In fact M_T is then an R_T-module with $\frac{r}{t} \cdot \frac{m}{t'} = \frac{rm}{tt'}$. There is a natural map $i: M \to M_T$ given by $i(m) = \frac{m}{1}$.

What this says is that M changes to M_T by (1) killing everything in M that is killed by <u>any</u> element of T, (2) allowing (formal) division by elements of T (by (1) this division becomes unique).

A common example is to consider $T = R - \mathfrak{p}$ for a prime ideal $\mathfrak{p}$. Then we write $M_{\mathfrak{p}} = M_T$ (we can never get confused as to the meaning since $0 \in \mathfrak{p}$ and $1 \notin \mathfrak{p}$).

For example, if p is a prime integer and $T = \{n \in Z \mid p \nmid n\}$ then for any abelian torsion group G, G_T is isomorphic to the p-primary part of G since all torsion prime to p is killed by T and division by elements of T is already well-defined in a p-primary group.

Let R be a graded ring, M an R-module and T a multiplicatively closed subset of R with elements of non-zero degree. Then M_T is either 0 or it has non-trivial elements of arbitrarily high and low degree. For example, if $r \in R$ is a non-nilpotent element of degree $n \neq 0$ then we can let $T = \{1, r, r^2, \ldots\}$. If M is an R-module then M_T is the set of all φr^k where $\varphi \in M/\{x \mid xr^n = 0$ for some positive $n\}$ and k runs through all integers. Then if $\varphi \neq 0$, $\varphi r^k \neq 0$ in degree kn + degree φ which can be made arbitrarily high or low depending on the choice of large positive or large negative k.

One very important aspect has not yet been mentioned: if $f: M \to N$ is an R-homomorphism, then $f_T: M_T \to N_T$ is well-defined by setting $f_T(\frac{m}{t}) = \frac{f(m)}{t}$. Observe that if $t \in T$ is of odd degree, then $2t^2 = 0$, hence, in R_T, $2t = \frac{2t^2}{t} = 0$. Thus either t is of even degree or $t = -t$ in R_T, hence no sign need be inserted when commuting elements of T with f_T.

Thus localization is a functor from the category of R-modules to the category of R_T-modules. Moreover, it is an exact functor: if $M \xrightarrow{f} N \xrightarrow{g} P$ is exact then look at $M_T \xrightarrow{f_T} N_T \xrightarrow{g_T} P_T$. Clearly $g_T \circ f_T = (g \circ f)_T = (0)_T = 0$; on the other hand $g_T(\frac{n}{t}) = 0$ implies $t'g(n) = 0$ for some $t' \in T$. Thus $g(t'n) = 0$ so $t'n = f(m)$ for some $m \in M$, by exactness. But then $f_T\left(\frac{m}{t't}\right) = \frac{f(m)}{t't} = \frac{t'n}{t't} = \frac{n}{t}$ so exactness is preserved.

In particular, then, if C is a chain complex and an R-module with R-module differentials, then $H_*(C)$ is an R-module and C_T is a chain complex. Then we can form $H_*(C)_T$ and $H_*(C_T)$ and the exactness of localization says that these are equal; that is, taking homology and localizing commute.

One can then pass, with great care, to localizing a spectral sequence: if $\{E^r, d^r\}$ is a spectral sequence of R-modules, then, in fact by the previous, there is a spectral sequence $\{{}^TE^r, {}^Td^r\}$ with ${}^TE^r_{**} = E^r_{**T}$ and ${}^Td^r = d^r_T$. This is a spectral sequence of R_T-modules. ${}^TE^\infty_{**}$ is <u>not</u>, however, necessarily equal to $(E^\infty_{**})_T$.

<u>Example</u>: Let $R = Z[t]$, $\deg t = 2$ and let the E^2 - term be pictured as

4								k_4		...
3						k_3				
2				k_2						
1		k_1								
0			m_1		m_2		m_3		m_4	...
	0	1	2	3	4	5	6	7	8	...

where E^2 is Z-free with generators $k_n \in E^2_{n-1,n}$ and $m_n \in E^2_{2n,0}$, $n \geq 1$. Let R act by $tk_n = 0$, $tm_n = m_{n+1}$. Let the differentials be given by $d^{n+1}(m_n) = k_n$. Thus $E^\infty_{**} = 0$. Let $T = \{1, t, t^2, \ldots\}$. Then $E^\infty_{**T} = 0$.

But ${}^TE^2_{**}$ has Z-free generators $m_n \in {}^TE^2_{2n,0}$ for all integers n, and is 0 off the base. Thus ${}^TE^\infty_{**} = {}^TE^2_{**} \neq 0$.

If, however, R acts on the fibre, instead of on the base, then we do get the two E^∞ terms related.

<u>Theorem 5.17</u>: Let $\{E^r_{**}, d^r\}$ be a spectral sequence of rings (i.e. E^r is a ring and d^r a derivation) with $E^r_{p,q} = 0$ for $p < 0$. Let $R = E^2_{0*}$, and let $T \subset R$ be multiplicatively closed. Then ${}^TE^\infty \cong E^\infty_T$.

<u>Proof</u>: We first observe that $d^r_{p,q} = E^r_{p,q} \to E^r_{p-r,q+r-1}$ is necessarily 0 for $r > p$ since $E^r_{p,q} = 0$ for $p < 0$. Also ${}^T d^r_{p,q} = 0$ for $r > p$ since ${}^T E^r_{p,q} = 0$ for $p < 0$. (But note that $E^r_{p,q} = 0$ for $q < 0$ does <u>not</u> imply that ${}^T E^r_{p,q} = 0$ for $q < 0$.)

Define $\varphi: E^\infty_{**T} \to {}^T E^\infty_{**}$ as follows: Let $\frac{[m]}{t} \in E^\infty_{**T}$. If $t \in T_s$ and $[m] \in E^\infty_{p,q+s}$ then we may consider $m \in E^r_{p,q+s}$ and $d^r(m) = 0$ for all $r > p$. Then $\frac{m}{t} \in {}^T E^r_{p,q}$ and ${}^T d^r(\frac{m}{t}) = \frac{d^r(m)}{t} = 0$ for all $r > p$, hence $[\frac{m}{t}] \in {}^T E^\infty_{p,q}$. Let $\varphi(\frac{[m]}{t}) = [\frac{m}{t}]$. It is easy to check that φ is well-defined.

Assume $\varphi(\frac{[m]}{t}) = 0$. Then for some n, $\frac{m}{t} = {}^T d^n(\frac{m'}{t'})$. Thus $\frac{m}{t} = \frac{d^n(m')}{t'}$. Thus for some $t'' \in T$, $t''(t'm - td^n(m')) = 0$. Then $d^n(t''tm') = t''t'm$ so $[t''t'm] = 0$ in E^∞_{**}, thus $\frac{[m]}{t} = \frac{[t''t'm]}{t''t't} = 0$ in E^∞_{**T}. Thus φ is $1-1$.

Let $[\frac{m}{t}] \in {}^T E^\infty_{p,q}$ be represented by $\frac{m}{t} \in {}^T E^{p+1}_{p,q}$. Then $m \in E^{p+1}_{p,q+s}$ where $t \in T_s$. Thus $d^n(m) = 0$ for all $n > p$ so $[m] \in E^\infty_{p,q+s}$ and so $\frac{[m]}{t} \in E^\infty_{**T}$ and clearly $\varphi(\frac{[m]}{t}) = [\frac{m}{t}]$. Thus φ is onto.

φ is thus an isomorphism and the theorem is proved.

Clearly we get a dual statement about cohomology-type spectral sequences where $R = E_2^{*0}$. For example in the Serre Spectral Sequence, a fibre space

$$F \xrightarrow{i} E \xrightarrow{p} B$$

with B 1-connected yields a spectral sequence

$$H^*(B;H^*(F)) \Rightarrow H^*(E) .$$

$H^*(B)$ is a ring under cup products, $H^*(E)$ is an $H^*(B)$-module via p^* and the spectral sequence commutes with the action of $H^*(B)$. So for a multiplicatively closed subset $T \subset H^*(B)$ we get a spectral sequence

$$H^*(B;H^*(F))_T \Rightarrow H^*(E)_T .$$

For example, we can get the following theorem

Theorem 5.18: If $x \in H^n(X)$ is non-nilpotent then either

1) x trangresses non-trivially to $H^{n-1}(\Omega X)$ or

2) there exists some $y \in H^r(X)$ where $n \nmid r$ such that $yx^m \neq 0$ for all m .

Proof: Assume that 2) fails. Then we localize at $T = \{1,x,x^2,\dots\}$. We have that $H^r(X)_T = 0$ if $n \nmid r$, in particular for $0 < r < n$. Then localizing the spectral sequence for $\Omega X \to PX \to X$ at T we see that $\frac{x}{1} \otimes 1 \in H^n(X;H^0(\Omega X))_T = E_2^{n,0}{}_T$ cannot be an $(n-1)$ boundary since $H^r(X)_T = 0$ for $0 < r < n$.

But there is a homomorphism $E_r^{**} \to E_{r_T}^{**}$ taking $x \otimes 1 \in E_r^{n\,o}$ to $\frac{x}{1} \otimes 1 \in E_{r_T}^{n,o}$ so $x \otimes 1$ cannot be an (n-1) boundary, hence it must be an n-boundary non-trivially, so x transgresses non-trivially to $H^{n-1}(\Omega X)$.

Let us apply these techniques to the spectral sequence

$$H_*(\underline{K};\pi_*(\underline{S})) \Rightarrow H_*(\underline{S}) \quad .$$

As mentioned earlier transgression occurs only for the Hopf classes, which are known to be nilpotent. Thus applying the same techniques as above we get

<u>Theorem 5.19</u>: If $\theta \in \pi_n(\underline{S})$ is non-nilpotent then there exists $\varphi \in \pi_r(\underline{S})$ with $n \nmid r$ such that $\theta^m\varphi \neq 0$ for all m.

Both of these theorems can be proved without localization by using an induction argument, (cf. <u>Kahn</u> for Theorem 5.19) but the proof given here is very slick and at the same time makes it more clear what is going on. The reader may try to prove Theorem 5.18 directly and will observe that although the idea is simple the proof is complicated - the idea of the proof, however, is the <u>only</u> thing that shows up in the localized version. I believe that this indicates that graded localization may be able to appreciably shorten some difficult proofs and at the same time keep the central ideas from being hidden.

Among those elements of $\pi_*(\underline{S})$ currently known, there is no candidate for a non-nilpotent element. Indeed, M. G. Barratt has conjectured that every element is nilpotent. One method for studying this question is to assume that there is a non-nilpotent element $\theta \in \pi_n(\underline{S})$, $n > 0$, (as in Theorem 5.19) and localize whatever we can at $T = \{1, \theta, \theta^2, \ldots\}$.

We shall now localize the stable category $\mathcal{S}$. Observe that for objects X and Y, $\{X,Y\}_*$ is a graded $\pi_*(S)$-module: if $f: S^d X \to Y$ and $g: S^n \to S^o$ then $g \wedge f: S^{n+d} X \to Y$. Let $\mathcal{S}_T$ have the same objects as $\mathcal{S}$, but with morphisms $\{X,Y\}_{*T}$, a $\pi_*(\underline{S})_T$-module. In this category, the action of $\theta \in \pi_n(\underline{S})$ (actually $\frac{\theta}{1} \in \pi_n(\underline{S})_T$) is invertible. That is, there exists $\theta^{-1} \in \pi_{-n}(\underline{S})_T$ with $\theta\theta^{-1} = \theta^{-1}\theta = 1$.

By the exactness of localization, if $f \in \{X,Y\}_*$ we can form C_f and we get the usual triangles <u>still exact after localization</u>. In particular, if we form the mapping cone of θ, C_θ, we find that $\{C_\theta, C_\theta\}_* = 0$ hence $C_\theta \simeq *$, a one point space, in $\mathcal{S}_T$.

By property 5 of mapping cones, for any maps $X \xrightarrow{f} Y \xrightarrow{g} Z$ there is a naturally defined map $C_f \to C_{gf}$ whose mapping cone is homotopy equivalent to C_g. If $f: X \to Y$ is a map in $\mathcal{S}$, we can consider θf as a composite $S^n X \xrightarrow{\theta \wedge 1_X} S^o X = X \xrightarrow{f} Y$. $C_{\theta \wedge 1_X} \simeq C_\theta \wedge X$ so we have defined a map $C_\theta \wedge X \to C_{\theta f}$ with mapping cone C_f. But in C_T, $C_\theta \simeq *$, hence $C_\theta \wedge X \simeq *$. Thus the induced map $\alpha_f: C_{\theta f} \to C_f$ is an equivalence in $\mathcal{S}_T$; that is, in $\mathcal{S}_T$ there is a morphism $\beta_f: C_f \to C_{\theta f}$ such that $\beta_f \alpha_f$ and $\alpha_f \beta_f$ are the identities.

Furthermore the following diagram commutes.

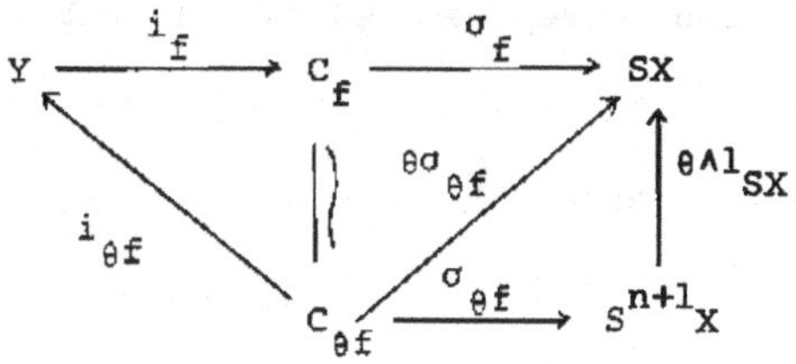

Thus up to $\mathcal{S}_T$ isomorphism we could say that given $C_{\theta f}$, $i_{\theta f}$ and $\sigma_{\theta f}$, we define C_f, i_f, σ_f by $C_f = C_{\theta f}, i_f = i_{\theta f}$ and $\sigma_f = \theta\sigma_{\theta f}$.

But now this allows us to define the mapping cone of <u>any</u> morphism of $\mathcal{S}_T$: whenever C_g, i_g, σ_g are defined, we define $C_{\frac{g}{\theta}} = C_g$, $i_{\frac{g}{\theta}} = i_g$, and $\sigma_{\frac{g}{\theta}} = \theta\sigma_g$ and this agrees with what we had before in the case $g = \theta f$ as morphisms of $\mathcal{S}$. Thus the category $\mathcal{S}_T$ itself has mapping cones, hence most of the usual properties of $\mathcal{S}$. (One thing we do lose, however, is ordinary homology. θ acts on homology trivially so the homology of any object in $\mathcal{S}$ is zero. Given any extraordinary homology theory where all powers of θ act non-trivially (for example stable homotopy) we would get a non-trivial theory on $\mathcal{S}_T$ by localizing).

Because of the existence of mapping cones, we can define Toda brackets. By the localization of Theorem 5.12, we find that θ itself will have a $\mathcal{S}_T$-Toda bracket representation in which non-trivial elements of degree > 0 and $< n$ appear. Translating this statement back to $\mathcal{S}$ and "stabilizing" (i.e. multiplying by a sufficiently high power of θ) we get:

<u>Theorem 5.20</u>: If $\theta \in \pi_n(\underline{S})$ is non-nilpotent, then for sufficiently large N, θ^N can be represented as a Toda bracket with entries φ such that

1) $n \nmid \deg \varphi$ and
2) $\varphi\theta^r \neq 0$ for all r

and all other entries must be of degree 0.

Then a possible line of attack on the question of the nilpotence of elements of $\pi_*(\underline{S})$ is to generalize the techniques of <u>Toda</u> [4]: A reasonable candidate for a non-nilpotent element was $\beta_1 \in \pi_{10}(\underline{S})$ of order 3. By Theorem 5.20, then we look at the Toda bracket representation of $\beta_1 = \langle\alpha_1, \alpha_1, \alpha_1\rangle$ and find that a reasonable candidate for φ is $\alpha_1 \in \pi_3(\underline{S})$ also of order 3 (although it is not necessarily the only candidate). But by a construction that makes use of the relation between β_1 and α_1 Toda proves $\beta_1^3\alpha_1 = 0$ leading one to suspect the nilpotence of β_1 and in fact $\beta_1^6 = 0$ follows almost immediately.

Using Theorem 5.20, then, a statement equivalent to the nilpotence of elements of $\pi_*(\underline{S})$ is the following

<u>Conjecture</u>: If φ appears in the Toda bracket representation of θ^r, for any r, and $\deg \theta \nmid \deg \varphi$ then some power of θ annihilates φ.

A study of this conjecture will probably be difficult and require tools at least as sophisticated as those of <u>Toda</u> [4], but this may be a partial step along the road to proving Barratt's conjecture.

BIBLIOGRAPHY

J. F. Adams, 1. On the non-existence of elements of Hopf invariant one, Ann. of Math. 72 (1960), 20-103.

2. Stable homotopy theory, Springer-Verlag, Berlin, 1966.

3. On the Groups J(X) - IV. Topology 5 (1966), 21-71.

R. Bott, A report on the unitary group, Proceedings of Symposia in Pure Mathematics, Vol. III, American Math. Soc., Providence, R. I., 1961, 1-6.

E. H. Brown, Jr., Cohomology theories, Ann. of Math., 75 (1962), 467-484.

H. Cartan, et al, Algèbres d'Eilenberg-MacLane et homotopie, Seminaire Henri Cartan de L'Ecole Normale Supérieure, 7e année (1954/55), 2 eme ed., Secretariat mathématique, 11 rue Pierre Curie, 1956.

J. M. Cohen, 1. Some results on the stable homotopy groups of Spheres, Bull. A. M. S. 72 (1966), 732-735.

2. The decomposition of stable homotopy, Ann. of Math. 87 (1968), 305-320.

3. A spectral sequence automorphism theorem; applications to fibre spaces and stable homotopy, Topology 7 (1968), 173-177.

A. Dold, Relations between ordinary and extraordinary homology, Coll. on Alg. Top., Aarhus (1962), 1-9.

P. Freyd, 1. Abelian categories, Harper and Row, New York, 1964.

2. Stable homotopy, Proc. of the Conf. on Cat. Alg., La Jolla, Springer, Verlag, Berlin, 1966.

3. The Grothendieck group for stable homotopy is free, Bull. A. M. S. 73 (1967), 84-86.

S. T. Hu, 1. Homotopy theory, Academic Press, New York, 1959.

2. Homology theory, Holden-Day, San Francisco, 1966.

D. W. Kahn, The spectral sequence of a Postnikov system, Comm. Math. Helv. 40 (1966), 169-198.

R. Kultze, Multiplikative eigenschaften von spektralen sequenzen, Math. Ann. 158 (1965), 233-267.

S. MacLane, Homology, Springer-Verlag, Berlin, 1963.

J. W. Milnor, On spaces having the homotopy type of a CW complex, Trans. A. M. S. 90 (1959), 272-280.

R. Mosher and M. Tangora, Cohomology operations, Harper and Row, New York, 1968.

E. H. Spanier, Algebraic topology, McGraw-Hill, New York, 1966.

N. Steenrod and D. B. A. Epstein, Cohomology operations, Princeton University Press, Princeton, 1962.

M. Tierney, Categorical constructions in stable homotopy theory, Springer-Verlag, Berlin, 1966.

H. Toda, 1. p-primary components of homotopy groups, IV, composition and toric constructions, Memoirs of the College of Science, Univ. of Kyoto, 32 (1959), 297-332.

2. Composition methods in homotopy groups of spheres, Princeton Univ. Press, 1962.

3. An important relation in homotopy groups of spheres, Proc. Japan Acad. 43 (1967), 839-842.

4. Extended p-th powers of complexes and applications to homotopy theory, Proc. Japan Acad. 44 (1968), 198-203.

K. Varadarajan, Groups for which Moore spaces $M(\pi,1)$ exist, Ann. of Math. 84 (1966), 368-371.

G. W. Whitehead, 1. Homotopy groups of joins and unions, Trans. A. M. S. 83 (1956), 55-69.

2. Generalized homology theory, Trans. A. M. S. 102 (1962), 227-283.

Lecture Notes in Mathematics

Bisher erschienen/Already published

Vol. 1: J. Wermer, Seminar über Funktionen-Algebren. IV, 30 Seiten. 1964. DM 3,80 / $ 1.10

Vol. 2: A. Borel, Cohomologie des espaces localement compacts d'après. J. Leray. IV, 93 pages. 1964. DM 9,– / $ 2.60

Vol. 3: J. F. Adams, Stable Homotopy Theory. Third edition IV, 78 pages. 1969. DM 8,– / $ 2.20

Vol. 4: M. Arkowitz and C. R. Curjel, Groups of Homotopy Classes. 2nd. revised edition. IV, 36 pages. 1967. DM 4,80 / $ 1.40

Vol. 5: J.-P. Serre, Cohomologie Galoisienne. Troisième édition. VIII, 214 pages. 1965. DM 18,– / $ 5.00

Vol. 6: H. Hermes, Term Logic with Choise Operator. III, 55 pages. 1970. DM 6,– / $ 1.70

Vol. 7: Ph. Tondeur, Introduction to Lie Groups and Transformation Groups. Second edition. VIII, 176 pages. 1969. DM 14,– / $ 3.80

Vol. 8: G. Fichera, Linear Elliptic Differential Systems and Eigenvalue Problems. IV, 176 pages. 1965. DM 13,50 / $ 3.80

Vol. 9: P. L. Ivănescu, Pseudo-Boolean Programming and Applications. IV, 50 pages. 1965. DM 4,80 / $ 1.40

Vol. 10: H. Lüneburg, Die Suzukigruppen und ihre Geometrien. VI, 111 Seiten. 1965. DM 8,– / $ 2.20

Vol. 11: J.-P. Serre, Algèbre Locale. Multiplicités. Rédigé par P. Gabriel. Seconde édition. VIII, 192 pages. 1965. DM 12,– / $ 3.30

Vol. 12: A. Dold, Halbexakte Homotopiefunktoren. II, 157 Seiten. 1966. DM 12,– / $ 3.30

Vol. 13: E. Thomas, Seminar on Fiber Spaces. IV, 45 pages. 1966. DM 4,80 / $ 1.40

Vol. 14: H. Werner, Vorlesung über Approximationstheorie. IV, 184 Seiten und 12 Seiten Anhang. 1966. DM 14,– / $ 3.90

Vol. 15: F. Oort, Commutative Group Schemes. VI, 133 pages. 1966. DM 9,80 / $ 2.70

Vol. 16: J. Pfanzagl and W. Pierlo, Compact Systems of Sets. IV, 48 pages. 1966. DM 5,80 / $ 1.60

Vol. 17: C. Müller, Spherical Harmonics. IV, 46 pages. 1966. DM 5,– / $ 1.40

Vol. 18: H.-B. Brinkmann und D. Puppe, Kategorien und Funktoren. XII, 107 Seiten, 1966. DM 8,– / $ 2.20

Vol. 19: G. Stolzenberg, Volumes, Limits and Extensions of Analytic Varieties. IV, 45 pages. 1966. DM 5,40 / $ 1.50

Vol. 20: R. Hartshorne, Residues and Duality. VIII, 423 pages. 1966. DM 20,– / $ 5.50

Vol. 21: Seminar on Complex Multiplication. By A. Borel, S. Chowla, C. S. Herz, K. Iwasawa, J.-P. Serre. IV, 102 pages. 1966. DM 8,– / $ 2.20

Vol. 22: H. Bauer, Harmonische Räume und ihre Potentialtheorie. IV, 175 Seiten. 1966. DM 14,– / $ 3.90

Vol. 23: P. L. Ivănescu and S. Rudeanu, Pseudo-Boolean Methods for Bivalent Programming. 120 pages. 1966. DM 10,– / $ 2.80

Vol. 24: J. Lambek, Completions of Categories. IV, 69 pages. 1966. DM 6,80 / $ 1.90

Vol. 25: R. Narasimhan, Introduction to the Theory of Analytic Spaces. IV, 143 pages. 1966. DM 10,– / $ 2.80

Vol. 26: P.-A. Meyer, Processus de Markov. IV, 190 pages. 1967. DM 15,– / $ 4.20

Vol. 27: H. P. Künzi und S. T. Tan, Lineare Optimierung großer Systeme. VI, 121 Seiten. 1966. DM 12,– / $ 3.30

Vol. 28: P. E. Conner and E. E. Floyd, The Relation of Cobordism to K-Theories. VIII, 112 pages. 1966. DM 9,80 / $ 2.70

Vol. 29: K. Chandrasekharan, Einführung in die Analytische Zahlentheorie. VI, 199 Seiten. 1966. DM 16,80 / $ 4.70

Vol. 30: A. Frölicher and W. Bucher, Calculus in Vector Spaces without Norm. X, 146 pages. 1966. DM 12,– / $ 3.30

Vol. 31: Symposium on Probability Methods in Analysis. Chairman. D. A. Kappos.IV, 329 pages. 1967. DM 20,– / $ 5.50

Vol. 32: M. André, Méthode Simpliciale en Algèbre Homologique et Algèbre Commutative. IV, 122 pages. 1967. DM 12,– / $ 3.30

Vol. 33: G. I. Targonski, Seminar on Functional Operators and Equations. IV, 110 pages. 1967. DM 10,– / $ 2.80

Vol. 34: G. E. Bredon, Equivariant Cohomology Theories. VI, 64 pages. 1967. DM 6,80 / $ 1.90

Vol. 35: N. P. Bhatia and G. P. Szegö, Dynamical Systems. Stability Theory and Applications. VI, 416 pages. 1967. DM 24,– / $ 6.60

Vol. 36: A. Borel, Topics in the Homology Theory of Fibre Bundles. VI, 95 pages. 1967. DM 9,– / $ 2.50

Vol. 37: R. B. Jensen, Modelle der Mengenlehre. X, 176 Seiten. 196 DM 14,– / $ 3.90

Vol. 38: R. Berger, R. Kiehl, E. Kunz und H.-J. Nastold, Differenti rechnung in der analytischen Geometrie IV, 134 Seiten. 1967 DM 12,– / $ 3.30

Vol. 39: Séminaire de Probabilités I. II, 189 pages. 1967. DM 14,– / $ 3.

Vol. 40: J. Tits, Tabellen zu den einfachen Lie Gruppen und ihren D stellungen. VI, 53 Seiten. 1967. DM 6.80 / $ 1.90

Vol. 41: A. Grothendieck, Local Cohomology. VI, 106 pages. 19 DM 10,– / $ 2.80

Vol. 42: J. F. Berglund and K. H. Hofmann, Compact Semitopologi Semigroups and Weakly Almost Periodic Functions. VI, 160 pag 1967. DM 12,– / $ 3.30

Vol. 43: D. G. Quillen, Homotopical Algebra. VI, 157 pages. 19 DM 14,– / $ 3.90

Vol. 44: K. Urbanik, Lectures on Prediction Theory. IV, 50 pages. 19 DM 5,80 / $ 1.60

Vol. 45: A. Wilansky, Topics in Functional Analysis. VI, 102 pages. 19 DM 9,60 / $ 2.70

Vol. 46: P. E. Conner, Seminar on Periodic Maps. IV, 116 pages. 19 DM 10,60 / $ 3.00

Vol. 47: Reports of the Midwest Category Seminar I. IV, 181 pages. 19 DM 14,80 / $ 4.10

Vol. 48: G. de Rham, S. Maumary et M. A. Kervaire, Torsion et Type Sim d'Homotopie. IV, 101 pages. 1967. DM 9,60 / $ 2.70

Vol. 49: C. Faith, Lectures on Injective Modules and Quotient Rin XVI, 140 pages. 1967. DM 12,80 / $ 3.60

Vol. 50: L. Zalcman, Analytic Capacity and Rational Approximation. 155 pages. 1968. DM 13.20 / $ 3.70

Vol. 51: Séminaire de Probabilités II. IV, 199 pages. 1968. DM 14,– / $ 3.90

Vol. 52: D. J. Simms, Lie Groups and Quantum Mechanics. IV, 90 pag 1968. DM 8,– / $ 2.20

Vol. 53: J. Cerf, Sur les difféomorphismes de la sphère de dimens trois (Γ_4 = O). XII, 133 pages. 1968. DM 12,– / $ 3.30

Vol. 54: G. Shimura, Automorphic Functions and Number Theory. 69 pages. 1968. DM 8,– / $ 2.20

Vol. 55: D. Gromoll, W. Klingenberg und W. Meyer, Riemannsche G metrie im Großen. VI, 287 Seiten. 1968. DM 20,– / $ 5.50

Vol. 56: K. Floret und J. Wloka, Einführung in die Theorie der lokalk vexen Räume. VIII, 194 Seiten. 1968. DM 16,– / $ 4.40

Vol. 57: F. Hirzebruch und K. H. Mayer, O (n)-Mannigfaltigkeiten, ex sche Sphären und Singularitäten. IV, 132 Seiten. 1968. DM 10,80 / $ 3

Vol. 58: Kuramochi Boundaries of Riemann Surfaces. IV, 102 pag 1968. DM 9,60 / $ 2.70

Vol. 59: K. Jänich, Differenzierbare G-Mannigfaltigkeiten. VI, 89 Sei 1968. DM 8,– / $ 2.20

Vol. 60: Seminar on Differential Equations and Dynamical Syste Edited by G. S. Jones. VI, 106 pages. 1968. DM 9,60 / $ 2.70

Vol. 61: Reports of the Midwest Category Seminar II. IV, 91 pages. 19 DM 9,60 / $ 2.70

Vol. 62: Harish-Chandra, Automorphic Forms on Semisimple Lie Gro X, 138 pages. 1968. DM 14,– / $ 3.90

Vol. 63: F. Albrecht, Topics in Control Theory. IV, 65 pages. 19 DM 6,80 / $ 1.90

Vol. 64: H. Berens, Interpolationsmethoden zur Behandlung von Ap ximationsprozessen auf Banachräumen. VI, 90 Seiten. 1968. DM 8,– / $ 2.20

Vol. 65: D. Kölzow, Differentiation von Maßen. XII, 102 Seiten. 19 DM 8,– / $ 2.20

Vol. 66: D. Ferus, Totale Absolutkrümmung in Differentialgeome und -topologie. VI, 85 Seiten. 1968. DM 8,– / $ 2.20

Vol. 67: F. Kamber and P. Tondeur, Flat Manifolds. IV, 53 pages. 19 DM 5,80 / $ 1.60

Vol. 68: N. Boboc et P. Mustaţă, Espaces harmoniques associés opérateurs différentiels linéaires du second ordre de type elliptic VI, 95 pages. 1968. DM 8,60 / $ 2.40

Vol. 69: Seminar über Potentialtheorie. Herausgegeben von H. Ba VI, 180 Seiten. 1968. DM 14,80 / $ 4.10

Vol. 70: Proceedings of the Summer School in Logic. Edited by M. H. L IV, 331 pages. 1968. DM 20,– / $ 5.50

Vol. 71: Séminaire Pierre Lelong (Analyse), Année 1967 – 1968. 19 pages. 1968. DM 14,– / $ 3.90

Bitte wenden / Contin

72: The Syntax and Semantics of Infinitary Languages. Edited by rwise. IV, 268 pages. 1968. DM 18,– / $ 5.00

73: P. E. Conner, Lectures on the Action of a Finite Group. IV, ages. 1968. DM 10,– / $ 2.80

74: A. Fröhlich, Formal Groups. IV, 140 pages. 1968. DM 12,– / $ 3.30

75: G. Lumer, Algèbres de fonctions et espaces de Hardy. VI, ages. 1968. DM 8,– / $ 2.20

76: R. G. Swan, Algebraic K-Theory. IV, 262 pages. 1968. 8,– / $ 5.00

77: P.-A. Meyer, Processus de Markov: la frontière de Martin. IV, ages. 1968. DM 10,– / $ 2.80

78: H. Herrlich, Topologische Reflexionen und Coreflexionen. XVI, Seiten. 1968. DM 12,– / $ 3.30

79: A. Grothendieck, Catégories Cofibrées Additives et Complexe ngent Relatif. IV, 167 pages. 1968. DM 12,– / $ 3.30

80: Seminar on Triples and Categorical Homology Theory. Edited . Eckmann. IV, 398 pages. 1969. DM 20,– / $ 5.50

81: J.-P. Eckmann et M. Guenin, Méthodes Algébriques en Méca- e Statistique. VI, 131 pages. 1969. DM 12,– / $ 3.30

82: J. Wloka, Grundräume und verallgemeinerte Funktionen. VIII, Seiten. 1969. DM 12,– / $ 3.30

83: O. Zariski, An Introduction to the Theory of Algebraic Surfaces.)0 pages. 1969. DM 8,– / $ 2.20

84: H. Lüneburg, Transitive Erweiterungen endlicher Permutations- ppen. IV, 119 Seiten. 1969. DM 10,– / $ 2.80

85: P. Cartier et D. Foata, Problèmes combinatoires de commu- n et réarrangements. IV, 88 pages. 1969. DM 8,– / $ 2.20

86: Category Theory, Homology Theory and their Applications I. ed by P. Hilton. VI, 216 pages. 1969. DM 16,– / $ 4.40

87: M. Tierney, Categorical Constructions in Stable Homotopy ory. IV, 65 pages. 1969. DM 6,– / $ 1.70

88: Séminaire de Probabilités III. IV, 229 pages. 1969. 18,– / $ 5.00

89: Probability and Information Theory. Edited by M. Behara, rickeberg and J. Wolfowitz. IV, 256 pages. 1969. DM 18,– / $ 5.00

90: N. P. Bhatia and O. Hajek, Local Semi-Dynamical Systems. 7 pages. 1969. DM 14,– / $ 3.90

91: N. N. Janenko, Die Zwischenschrittmethode zur Lösung mehr- ensionaler Probleme der mathematischen Physik. VIII, 194 Seiten. . DM 16,80 / $ 4.70

92: Category Theory, Homology Theory and their Applications II. ed by P. Hilton. V, 308 pages. 1969. DM 20,– / $ 5.50

93: K. R. Parthasarathy, Multipliers on Locally Compact Groups. 4 pages. 1969. DM 5,60 / $ 1.60

94: M. Machover and J. Hirschfeld, Lectures on Non-Standard ysis. VI, 79 pages. 1969. DM 6,– / $ 1.70

95: A. S. Troelstra, Principles of Intuitionism. II, 111 pages. 1969. 10,– / $ 2.80

96: H.-B. Brinkmann und D. Puppe, Abelsche und exakte Kate- en, Korrespondenzen. V, 141 Seiten. 1969. DM 10,– / $ 2.80

97: S. O. Chase and M. E. Sweedler, Hopf Algebras and Galois ory. II, 133 pages. 1969. DM 10,– / $ 2.80

98: M. Heins, Hardy Classes on Riemann Surfaces. III, 106 pages. . DM 10,– / $ 2.80

99: Category Theory, Homology Theory and their Applications III. ed by P. Hilton. IV, 489 pages. 1969. DM 24,– / $ 6.60

100: M. Artin and B. Mazur, Etale Homotopy. II, 196 Seiten. 1969. 12,– / $ 3.30

101: G. P. Szegö et G. Treccani, Semigruppi di Trasformazioni tivoche. VI, 177 pages. 1969. DM 14,– / $ 3.90

102: F. Stummel, Rand- und Eigenwertaufgaben in Sobolewschen men. VIII, 386 Seiten. 1969. DM 20,– / $ 5.50

103: Lectures in Modern Analysis and Applications I. Edited by . Taam. VII, 162 pages. 1969. DM 12,– / $ 3.30

104: G. H. Pimbley, Jr., Eigenfunction Branches of Nonlinear rators and their Bifurcations. II, 128 pages. 1969. DM 10,– / $ 2.80

105: R. Larsen, The Multiplier Problem. VII, 284 pages. 1969. 18,– / $ 5.00

106: Reports of the Midwest Category Seminar III. Edited by S. Mac e. III, 247 pages. 1969. DM 16,– / $ 4.40

107: A. Peyerimhoff, Lectures on Summability. III, 111 pages. . DM 8,– / $ 2.20

108: Algebraic K-Theory and its Geometric Applications. Edited . M.F. Moss and C. B. Thomas. IV, 86 pages. 1969. DM 6,– / $ 1.70

Vol. 109: Conference on the Numerical Solution of Differential Equations. Edited by J. Ll. Morris. VI, 275 pages. 1969. DM 18,– / $ 5.00

Vol. 110: The Many Facets of Graph Theory. Edited by G. Chartrand and S. F. Kapoor. VIII, 290 pages. 1969. DM 18,– / $ 5.00

Vol. 111: K. H. Mayer, Relationen zwischen charakteristischen Zahlen. III, 99 Seiten. 1969. DM 8,– / $ 2.20

Vol. 112: Colloquium on Methods of Optimization. Edited by N. N. Moiseev. IV, 293 pages. 1970. DM 18,– / $ 5.00

Vol. 113: R. Wille, Kongruenzklassengeometrien. III, 99 Seiten. 1970. DM 8,– / $ 2.20

Vol. 114: H. Jacquet and R. P. Langlands, Automorphic Forms on GL (2). VII, 548 pages. 1970. DM 24,– / $ 6.60

Vol. 115: K. H. Roggenkamp and V. Huber-Dyson, Lattices over Orders I. XIX, 290 pages. 1970. DM 18,– / $ 5.00

Vol. 116: Sèminaire Pierre Lelong (Analyse) Année 1969. IV, 195 pages. 1970. DM 14,– / $ 3.90

Vol. 117: Y. Meyer, Nombres de Pisot, Nombres de Salem et Analyse Harmonique. 63 pages. 1970. DM 6.– / $ 1.70

Vol. 118: Proceedings of the 15th Scandinavian Congress, Oslo 1968. Edited by K. E. Aubert and W. Ljunggren. IV, 162 pages. 1970. DM 12,– / $ 3.30

Vol. 119: M. Raynaud, Faisceaux amples sur les schémas en groupes et les espaces homogènes. III, 219 pages. 1970. DM 14,– / $ 3.90

Vol. 120: D. Siefkes, Büchi's Monadic Second Order Successor Arithmetic. XII, 130 Seiten. 1970. DM 12,– / $ 3.30

Vol. 121: H. S. Bear, Lectures on Gleason Parts. III, 47 pages. 1970. DM 6,–/$ 1.70

Vol. 122: H. Zieschang, E. Vogt und H.-D. Coldewey, Flächen und ebene diskontinuierliche Gruppen. VIII, 203 Seiten. 1970. DM 16,– / $ 4.40

Vol. 123: A. V. Jategaonkar, Left Principal Ideal Rings. VI, 145 pages. 1970. DM 12,– / $ 3.30

Vol. 124: Séminare de Probabilités IV. Edited by P. A. Meyer. IV, 282 pages. 1970. DM 20,– / $ 5.50

Vol. 125: Symposium on Automatic Demonstration. V, 310 pages. 1970. DM 20,– / $ 5.50

Vol. 126: P. Schapira, Théorie des Hyperfonctions. XI, 157 pages. 1970. DM 14,– / $ 3.90

Vol. 127: I. Stewart, Lie Algebras. IV, 97 pages. 1970. DM 10,– / $ 2.80

Vol. 128: M. Takesaki, Tomita's Theory of Modular Hilbert Algebras and its Applications. II, 123 pages. 1970. DM 10,– / $ 2.80

Vol. 129: K. H. Hofmann, The Duality of Compact Semigroups and C*- Bigebras. XII, 142 pages. 1970. DM 14,– / $ 3.90

Vol. 130: F. Lorenz, Quadratische Formen über Körpern. II, 77 Seiten. 1970. DM 8,– / $ 2.20

Vol. 131: A Borel et al., Seminar on Algebraic Groups and Related Finite Groups. VII, 321 pages. 1970. DM 22,– / $ 6.10

Vol. 132: Symposium on Optimization. III, 348 pages. 1970. DM 22,– / $ 6.10

Vol. 133: F. Topsøe, Topology and Measure. XIV, 79 pages. 1970. DM 8,– / $ 2.20

Vol. 134: L. Smith, Lectures on the Eilenberg-Moore Spectral Sequence. VII, 142 pages. 1970. DM 14,– / $ 3.90

Vol. 135: W. Stoll, Value Distribution of Holomorphic Maps into Compact Complex Manifolds. II, 267 pages. 1970. DM 18,– / $

Vol. 136: M. Karoubi et al., Séminaire Heidelberg-Saarbrücken-Strasbuorg sur la K-Théorie. IV, 264 pages. 1970. DM 18,– / $ 5.00

Vol. 137: Reports of the Midwest Category Seminar IV. Edited by S. MacLane. III, 139 pages. 1970. DM 12,– / $ 3.30

Vol. 138: D. Foata et M. Schützenberger, Théorie Géométrique des Polynômes Eulériens. V, 94 pages. 1970. DM 10,– / $ 2.80

Vol. 139: A. Badrikian, Séminaire sur les Fonctions Aléatoires Linéaires et les Mesures Cylindriques. VII, 221 pages. 1970. DM 18,– / $ 5.00

Vol. 140: Lectures in Modern Analysis and Applications II. Edited by C. T. Taam. VI, 119 pages. 1970. DM 10,– / $ 2.80

Vol. 141: G. Jameson, Ordered Linear Spaces. XV, 194 pages. 1970. DM 16,– / $ 4.40

Vol. 142: K. W. Roggenkamp, Lattices over Orders II. V, 388 pages. 1970. DM 22,– / $ 6.10

Vol. 143: K. W. Gruenberg, Cohomological Topics in Group Theory. XIV, 275 pages. 1970. DM 20,– / $ 5.50